## More Praise for
### *Parenting from the Inside Out*

'*Parenting from the Inside Out* shows parents how to understand and integrate the experiences from their own childhood into a nurturing style of parenting that promotes healthy communication, meaningful attachment, and trust.'
—SAL SEVERE, PhD, author of *How to Behave So Your Children Will, Too!*

'This is an excellent book, with a positive, humane, and compassionate message. I applaud the emphasis on emotional connection and communication between parent and child and the thesis that as you understand and connect with yourself, you can better connect with your child. This book will be deeply interesting to parents, as well as useful for them.'
—L. ALAN SROUFE, PhD, professor, Institute of Child Development, University of Minnesota, Minneapolis, and author of *Emotional Development: the organization of emotional life in the early years*

'Daniel Siegel and Mary Hartzell's illuminating and inspiring book presents a practical model of parenting that nurtures an empathic dialogue within oneself and with one's children. Consequently, parenting becomes not just something that we *do* but something that is very much a part of who we *are*. As a parent of seven children, I've given this book a permanent spot on my night table.'
—KATE CAPSHAW SPIELBERG

'This is must reading for all parents. Parenting is the most important job in the world, and this book makes it more understandable and easier.'
—HAROLD S. KOPLEWICZ, MD, president, Child Mind Institute, and author of *More than Moody: recognizing and treating adolescent depression*

'A new and thought-provoking approach to parenting.'
—ELLEN GALINSKY, president, Families and Work Institute and author of *Ask the Children: the breakthrough study that reveals how to succeed at work and parenting*

'Every parent should read *Parenting from the Inside Out*. Full of fascinating information about brain development that is not readily available in any other parenting books, this book will make watching your children learn and grow even more interesting than it already is.'

—BETTY EDWARDS, author of *Drawing on the Right Side of the Brain*

'Anyone who feels doomed to repeat his parents' errors should pick up a copy of this book. Readers learn to explore and resolve issues from the past, freeing them to be their best selves, flexible and fully engaged.'

—JESSICA TEICH, author of *Trees Make the Best Mobiles:*
*simple ways to raise your child in a complex world*

'*Parenting from the Inside Out* is an extraordinary tool for parenting. The gift of the book is that it is a beautiful reminder to be more compassionate with our children and, equally important, with ourselves. It reveals that it's never too late to heal your old wounds and re-parent yourself. By doing this, you are allowed to be more present and joyful in your child's journey. Parent and child both emerge more whole and more loving.'

—JESSIE NELSON, director, producer, and screenwriter

'The unique structure of this book gives us a rare opportunity to reflect on the parenting we received as children, to keep what works, and to re-think how we would like to parent our own children. These fascinating insights can alter the way we think about our own past experiences and their effects on who we are today.'

—SIR RICHARD BOWLBY, BT, chair of trustees,
Centre for Attachment-Based Psychoanalytic Psychotherapy

'This book uniquely pairs the clinical experience of a psychiatrist with the deep wisdom of a nursery-school teacher whose daily practice is enmeshed in the minute-to-minute frustrations, joys, despondency, and exhilaration that constitute the experience of every child. Together Siegel and Hartzell delicately peel back the many layers of parenting to reveal the pure nature of the relationship that is at its core.'

—NEAL HALFON, MD, MPH, professor of paediatrics,
community health sciences, and policy studies, UCLA; director,
UCLA Center for Healthier Children, Families and Communities;
and editor of *Child Rearing in America:*
*challenges facing parents with young children*

# PARENTING FROM THE INSIDE OUT

DANIEL J. SIEGEL, MD, received his medical degree from Harvard University and completed his postgraduate medical education at UCLA, where he is currently a clinical professor. He is the executive director of the Mindsight Institute, and the author of numerous books, including the acclaimed bestsellers *Mindsight: change your brain and your life* and *The Whole-Brain Child* (co-authored with Tina Payne Bryson). He lives in Los Angeles with his wife and occasionally with his launched adolescents.

MARY HARTZELL, MED, received her master's degree in early-childhood education and psychology from UCLA, and is a child-development specialist and parent educator. She has taught children, parents, and teachers for more than 30 years, and has been the director of a highly respected Reggio-inspired preschool in Santa Monica, California. Her parent-education classes and CD series on parent–child relationships have improved the lives of children and their families for decades.

# Parenting from the Inside Out

how a deeper self-understanding can
help you raise children who thrive

**TENTH-ANNIVERSARY EDITION**

## DANIEL J. SIEGEL, MD
## & MARY HARTZELL, MEd

SCRIBE
*Melbourne • London*

Scribe Publications
18–20 Edward St, Brunswick, Victoria 3056, Australia
2 John St, Clerkenwell, London, WC1N 2ES, United Kingdom

Published by Scribe 2014
Reprinted 2016 (twice), 2017

This edition published by arrangement with Jeremy P. Tarcher/Penguin

Book design by Gretchen Achilles

Printed and bound in the UK by CPI Group (UK) Ltd, Croydon CR0 4YY

Scribe Publications is committed to the sustainable use of natural resources and the use of paper products made responsibly from those resources.

9781922247445 (UK edition)
9781922070937 (Australian edition)
9781925113068 (e-book)

CiP records for this title are available from the National Library of Australia and the British Library

scribepublications.co.uk
scribepublications.com.au

To our children

For the joy and the wisdom they bring to our lives and

In appreciation of our parents

For the precious gift of life and for all that we have learned from them

# Contents

# Preface to the Tenth-Anniversary Edition

I t is a pleasure to welcome you to this tenth-anniversary edition of *Parenting from the Inside Out*. When we first put together this approach, we were inspired by recent scientific findings that revealed that the best predictor of a child's security of attachment to a caregiver is the way that adult has made sense of his or her own childhood experiences. We wanted to create a practical application of this important scientific discovery so that parents around the world could directly benefit from it. Over the past decade, more than ten thousand individuals and their children across a wide range of cultures and social and economic backgrounds were studied, and the findings of all of this research have further demonstrated this crucial parenting-from-the-inside-out principle: making sense of your life is the best gift you can give your child, or yourself.

Secure attachment is but one piece of a large developmental puzzle that includes many factors that influence how our children grow into their adolescent and adult years. While secure attachment supports the development of resilience and well-being in children, many other issues, such as genetics, peers, and experiences in school and our larger society, influence how our children turn out in life. Yet attachment is one factor we, as parents, can influence directly in our children's lives because of this crucial inside-out idea: it isn't what happened to you in your childhood that is the critical factor—it is how you make sense of how those experiences have influenced your life. Knowing this powerful scientific finding, we put together a step-by-step approach that parents and other caregivers can use to optimize not only their relationships with the children they care for but also their relationships with other adults as well. It's a win-win-win situation: your children will thrive, your interpersonal relationships will prosper, and even your relationship with yourself will blossom and become filled with more self-compassion.

Because we develop across the life span, taking this inside-out journey is helpful no matter your age. We've received enthusiastic feedback from young adults to people in their later years—even into their eighties and nineties! Because how our children are attached to us influences how they develop from their earliest years onward, it is never too late to make sense of your life to help your children, be they toddlers, adolescents, or even adults. We've thought of this approach as a scientifically inspired, practical strategy that delivers itself as one big, supportive hug. It's not always easy to make sense of your life, and so we've woven lots of nurturing, practical tips, and factual knowledge into this approach to help you along the way. That so many people from across the planet have told us that this is a "favorite" book for them suggests that the inside-out journey is worth the effort. You'll thrive, and your children will, too! What more can you ask for in focusing your emotional energy and your precious time in this life? Enjoy, and let us know how it all goes for you!

*Dan and Mary*

# Introduction

## AN INSIDE-OUT APPROACH TO PARENTING

How you make sense of your childhood experiences has a profound effect on how you parent your own children. In this book, we will reflect on how self-understanding influences the approach you bring to your role as a parent. Understanding more about yourself in a deeper way can help you build a more effective and enjoyable relationship with your children.

As we grow and understand ourselves we can offer a foundation of emotional well-being and security that enables our children to thrive. Research in the field of child development has demonstrated that a child's security of attachment to parents is very strongly connected to the parents' understanding of their own early life experiences. Contrary to what many people believe, your early experiences do not determine your fate. If you had a difficult childhood but have come to make sense of those experiences, you are not bound to re-create the same negative interactions with your own children. Without such self-understanding, however, science has shown that history will likely repeat itself, as negative patterns of family interactions are passed down through the generations. This book is designed to help you make sense of your own life, both past and present, by enhancing your understanding of how your childhood has influenced your life and affects your parenting.

When we become parents we are given an incredible opportunity to grow as individuals because we ourselves are put back into an intimate parent-child relationship, this time in a different role. So many times parents have said, "I never thought I'd do or say the very things to my children that felt hurtful to me when I was a child. And yet I find myself doing exactly that." Parents can feel stuck in repetitive, unproductive patterns that don't support the loving, nurturing relationships they envisioned when they began their roles as parents. Making sense of life can free parents from patterns of the past that have imprisoned them in the present.

## WHO WE ARE

The authors bring to this book their own distinct professional experiences in working with parents and children, Dan as a child psychiatrist and Mary as an early childhood and parent educator. Both are parents, Dan with children now beyond their teens, Mary with grown children and grandchildren.

Mary has worked as an educator with children and families for over forty years, directing a nursery school; teaching children, parents, and teachers; and consulting individually with parents. All of her experiences have provided her opportunities to share in the lives of families and to learn a great deal about the frustrations, and joys, parents experience in this sometimes daunting task. In developing her parent education course, Mary found that when parents had an opportunity to reflect on their own childhood experiences they could make more effective choices in raising their own children.

Mary and Dan first met when Dan's daughter attended the nursery school that Mary directed. The school's approach incorporates a profound respect for the child's emotional experience and fosters dignity and creativity among the children, parents, and teaching staff. During that time, Dan worked on the parent education committee and gave a few lectures on brain development for the parents and faculty. When

Mary and Dan realized how parallel their approaches to parenting were, they decided to work together to create an integrated seminar.

Dan's interest in the science of development along with his work as a child psychiatrist offered a different but complementary perspective. For more than two decades Dan has worked to synthesize a wide range of scientific disciplines into an integrated developmental framework for understanding the mind, the brain, and human relationships. This convergence of scientific perspectives has been called "interpersonal neurobiology" and is now being incorporated into a number of professional educational programs focusing on mental health and emotional well-being. When his book *The Developing Mind: Toward a Neurobiology of Interpersonal Experience* (Guilford Press, 1999/2012) was first published, Mary and Dan were teaching their integrated course. The enthusiastic feedback from parents inspired them to begin collaboration on this book. Many parents in the seminar mentioned how useful these ideas were in helping them to understand themselves and to make meaningful connections with their children. "Why don't you write something together so that others can get some of that exciting science and wisdom—and energy—that you brought to our seminar?" numerous parents urged.

We are thrilled to have the chance to share this approach with you. We hope this book conveys in an enjoyable and accessible way some of these practical ideas about the art and science of nurturing relationships and self-knowledge. It is our hope that this book will help you and your child find more joy in each other each day.

**BUILDING RELATIONSHIPS AND SELF-UNDERSTANDING**

The way we communicate with our children has a profound impact on how they develop. Our ability to have sensitive, reciprocal communication nurtures a child's sense of security, and these trusting secure relationships help children do well in many areas of their lives. Our ability to communicate effectively in creating security in our children is most strongly predicted by our having made sense of the events of our own early life. Making sense of our life enables us to understand and integrate our own childhood experiences, positive or negative, and to accept them as part of our ongoing life story. We can't change what happened to us as children but we can change the way we think about those events.

Thinking about our lives in a different way entails being aware of our present experiences, including our emotions and perceptions, and appreciating how the present is affected by events from the past. Understanding how we remember and how we construct a picture of ourselves as a part of the world we live in can help us to make sense of how the past continues to impact our lives. How does making sense of our lives help our children? By freeing ourselves from the constraints of our past, we can offer our children the spontaneous and connecting relationships that enable them to thrive. By deepening our ability to understand our own emotional experience, we are better able to relate empathically with our children and promote their self-understanding and healthy development.

In the absence of reflection, history often repeats itself, and parents are vulnerable to passing on to their children unhealthy patterns from the past. Understanding our lives can free us from the otherwise almost predictable situation in which we re-create the damage to our children that was done to us in our own childhoods. Research has clearly demonstrated that our children's attachment to us will be influenced by what happened to us when we were young if we do not come to process and understand those experiences. By making sense of our lives we can deepen a capacity for self-understanding and bring coherence to our emotional experience, our views of the world, and our interactions with our children.

Of course, the development of the child's whole personality is influenced by many things, including genetics, temperament, physical health, and experience. Parent-child relationships offer one very important part of the early experience that directly shapes a child's emerging personality. Emotional intelligence, self-esteem, cognitive abilities, and social skills are built on this early attachment relationship. How parents have reflected on their lives directly shapes the nature of that relationship.

Even if we have come to understand ourselves well, our children will still make their own journey through life. Although we may supply them with a secure foundation through our own deeper self-understanding, our role as parents serves to support our children's development, not guarantee its outcome. Research suggests that children who have had a positive connection in life have a source of resilience for dealing with life's challenges. Building a positive relationship with our children involves being open to our own growth and development.

Through the process of reflection, you can enhance the coherence of your life narrative and improve your relationship with your child. No one had a "perfect childhood" and some of us had more challenging experiences than others. Yet even those with overwhelmingly difficult past experiences can come to resolve those issues and have meaningful and rewarding relationships with their children. Research has shown the exciting finding that parents who themselves did not have "good enough parents" or who even had traumatic childhoods can make sense of their lives and have healthy relationships. More important for our children than merely what happened to us in the past is the way we have come to process and understand it. The opportunity to change and grow continues to be available throughout our lives.

## ABOUT THIS BOOK

This is not a how-to book—it is a "how we" book. We will explore new insights into parenting by examining such processes as how we remember, perceive, feel, communicate, attach, make sense, disconnect and reconnect, and reflect with our children on the nature of their internal experiences. We will explore issues from recent research on parent-child relationships and integrate these with new discoveries from brain science. By examining the science of how we experience and connect, a new perspective emerges that can help us deepen our understanding of ourselves, our children, and our relationships with each other.

The "Inside-Out Exercises" at the end of each chapter can be used to explore new possibilities for internal understanding and interpersonal communication. The reflections encouraged by these exercises have helped parents deepen their understanding of current and past experiences and enabled them to improve their relationships with their children.

You may find that writing your experiences of these exercises in a journal will provide an opportunity for deeper reflection and self-knowledge. Writing can help to free your mind from the entanglements of the past and to liberate you to better understand yourself. Your journal entries can include anything you would like, such as drawings, reflections, descriptions, and stories. Some people don't like to write at all and may prefer to reflect in solitude or to talk with a friend. Others like to write in a journal each day or when they feel motivated by an experience or an emotion. Do what feels right for you.

The value of the exercises will come from an open attitude and thoughtful reflections.

The shaded "Spotlight on Science" sections at the end of each chapter provide the interested reader with more information about various aspects of research findings that are relevant to parenting. These sections are not essential to understand and use the ideas in the main body of the book. They offer expanded, in-depth information but reader-friendly discussion that provides more on the scientific foundation for many of the concepts in the book and we hope you will find them thought provoking and useful. Each of the "Spotlight on Science" sections relates to the main text but also stands on its own and can be read independent of other Spotlight entries. Reading all of the Spotlight sections will offer an interdisciplinary view of aspects of science relevant to the process of parenting.

There is a saying in science that "chance favors the prepared mind." Having knowledge about the science of development and human experience can prepare your mind to build a deeper understanding of the emotional lives of your children and of you yourself. Whether you read the Spotlight sections after completing each chapter or choose to explore them after finishing the main text of the book, we hope that you will feel free to take an approach that respects your own unique style of learning.

## AN APPROACH TO PARENTING

This book will encourage you to build an approach to parenting that is founded on basic principles of internal understanding and interpersonal connection. The anchor points for this approach to the parent-child relationship are mindfulness, lifelong learning, response flexibility, mindsight, and joyful living.

## Being Mindful

Mindfulness is at the heart of nurturing relationships. When we are mindful, we live in the present moment and are aware of our own thoughts and feelings and also are open to those of our children. The ability to stay present with clarity within ourselves allows us to be fully present with others and to respect each person's individual experience. No two people see things in exactly the same way. Mindfulness gives respect to the individuality of each person's unique mind.

When we are being fully present as parents, when we are mindful, it enables our children to fully experience themselves in the moment. Children learn about themselves by the way we communicate with them. When we are preoccupied with the past or worried about the future, we are physically present with our children but are mentally absent. Children don't need us to be fully available all the time, but they do need our presence during connecting interactions. Being mindful as a parent means having intention in your actions. With intention, you purposefully choose your behavior with your child's emotional well-being in mind. Children can readily detect intention and thrive when there is purposeful interaction with their parents. It is within our children's emotional connections with us that they develop a deeper sense of themselves and a capacity for relating.

## Lifelong Learning

Your children give you the opportunity to grow and challenge you to examine issues left over from your own childhood. If you approach such challenges as a burden, parenting can become an unpleasant chore. If, on the other hand, you try to see these moments as learning opportunities, then you can continue to grow and develop. Having the attitude that you can learn throughout your life enables you to approach parenting with an open mind, as a journey of discovery.

The mind continues to develop throughout the life span. The mind emanates from the activity of the brain; thus, findings from brain science can inform our understanding of ourselves. Recent findings from neuroscience point to the exciting idea that the brain continues to develop both new connections and even new neurons in certain areas throughout a person's life. The connections among neurons determine how mental processes are created. Experience shapes neural connections in the brain. Therefore, experience shapes the mind. Interpersonal relationships and self-reflection foster the ongoing growth of the mind: being a parent offers us an opportunity to continue to learn as we reflect on our experiences from new and ever-evolving points of view. Parenting also gives us the opportunity to create an attitude of openness in our children as we nurture their curiosity and support their ongoing explorations of the world. The complex and often challenging interactions of parenting give us the opportunity to create new possibilities for the growth and development of our children and ourselves.

## Response Flexibility

Being able to respond in flexible ways is one of the biggest challenges of being a parent. Response flexibility is the ability of the mind to sort through a wide variety of mental processes, such as impulses, ideas, and feelings, and come up with a thoughtful, nonautomatic response. Rather than merely automatically reacting to a situation, an individual can reflect and intentionally choose an appropriate direction of action. Response flexibility is the opposite of a "knee-jerk reaction." It involves the capacity to delay gratification and to inhibit impulsive behaviors. This ability is a cornerstone of emotional maturity and compassionate relationships.

Under certain conditions, response flexibility may be impaired. When tired, hungry, frustrated, disappointed, or angered, we can lose the ability to be reflective and become limited in our capacity to choose our behaviors. We may be swept up in our own emotions and lose perspective. At these times, we can no longer think clearly and are at high risk of overreacting and causing distress to our children.

Children challenge us to remain flexible and to maintain emotional equilibrium. It can be difficult to balance flexibility with the importance of structure in a child's life. Parents can learn how to achieve this balance and nurture flexibility in their children by modeling flexible responses in their own interactions. When we are flexible, we have a choice about what behaviors to enact and what parental approach and values to support. We have the ability to be proactive and not just reactive. Response flexibility enables us to contain a wide array of emotions and to think through how we will respond after we consider another's point of view. When parents have the ability to respond with flexibility to their children, it is more likely that their children will develop flexibility as well.

## Mindsight

Mindsight is the ability to perceive our own minds and the minds of others. Our minds create representations of objects and ideas. For example, we can visualize the image of a flower or a dog in our minds, but there is no plant or animal in our heads, just a neurally constructed symbol containing information about that object. Mindsight depends upon the ability of the mind to create mental symbols of the mind itself. This ability allows us to focus on the thoughts, feelings, perceptions, sensations, memories, beliefs, attitudes, and intentions of others as well as of ourselves. These are the basic elements of the mind that we can perceive and use to understand our children and ourselves.

Parents often respond to their child's behavior by focusing on the surface level of the experience and not on the deeper level of the mind. At times we may look only at the behaviors of others and merely notice the ways people act without seeing the internal mental processes leading to their actions. However, there is a deeper level beneath behavior, and this is the root of motivation and action. This deeper level is the mind. Mindsight enables us to focus on more than just the surface level of experience. Parents who focus on the level of mind with their children nurture the development of emotional understanding and compassion. Talking with children about their thoughts, memories, and feelings provides them with the essential interpersonal experiences necessary for self-understanding and building their social skills.

Mindsight allows parents to "see" the minds of their children through the basic signals they can perceive. Verbal information, the words people use, are only a part of how we come to understand others. The nonverbal messages of eye contact, facial expression, tone of voice, gestures and touch, body posture, and the timing and intensity of response are also extremely important elements of communication. These nonverbal signals may reveal our internal processes more directly than our words. Being sensitive to nonverbal communication helps us to better understand our children and allows us to consider their point of view and to relate with compassion.

## Joyful Living

Enjoying your child and sharing in the awe of discovering what it means to be alive, to be a person in a wondrous world, is crucial for the development of your child's positive sense of self. When we are respectful and compassionate toward ourselves and our children, we often gain a fresh perspective that can enrich our enjoyment of life together. Remembering and reflecting on the experiences of day-to-day life creates a deep sense of feeling connected and understood.

Parents can accept their child's invitation to slow down and appreciate the beauty and connection that life offers each day. When parents feel pressure in their busy lives, they may often feel strained to keep up with all the details of managing family schedules. Children need to be enjoyed and valued, not managed. We often focus on the problems of life rather than on the possibilities for enjoyment and learning available to us. When we are too busy doing things for our children, we forget how important it is to simply be with them. We can delight in the opportunity to join with our children in the amazing experience of growing together. Learning to share in the joy of living is at the heart of a rewarding parent-child relationship.

When we become parents, we often see ourselves as our children's teachers, but we soon discover that our children are our teachers as well. Through this intimate relationship our past, present, and future take on new meaning as we share experiences and create memories that greatly enrich our lives together. We hope you will find this book beneficial in supporting your growth and development as a person and as a parent so that your ongoing relationship with your children may continue to deepen throughout your life.

# CHAPTER 1

# How We Remember: Experience Shapes Who We Are

## INTRODUCTION

When we become parents, we bring with us issues from our own past that influence the way we parent our children. Experiences that are not fully processed may create unresolved and leftover issues that influence how we react to our children. These issues can easily get triggered in the parent-child relationship. When this happens our responses toward our children often take the form of strong emotional reactions, impulsive behaviors, distortions in our perceptions, or sensations in our bodies. These intense states of mind impair our ability to think clearly and remain flexible and affect our interactions and relationships with our children. At these times, we're not acting like the parents we want to be and are often left wondering why this role of parenting sometimes seems to "bring out the worst in us." Issues that are rooted in our past impact our present reality and directly affect the way we experience and interact with our children even when we're not aware of their origins.

To our role of parenting we bring our own emotional baggage, which can unpredictably interfere in our relationship with our children. Leftover issues or unresolved trauma and loss involve significant themes from the past that stem from repeated experiences early in life that were difficult and emotionally significant. These issues, especially if we have not reflected on them and integrated them into our self-understanding, can continue to affect us in the present. For example, if your mother often left the house without saying good-bye because she didn't want to hear you cry, your sense of trust would be broken, especially at times of possible separation. You'd feel insecure and uncertain. After she'd leave without your awareness, you'd look for her and get upset at her absence. The situation would be made more stressful if the adult who was caring for you insisted that you not cry. Not only would you feel betrayed and distressed at the loss of your mother, but because no caring adult gave you comfort by listening to you, empathizing with your feelings, and giving you some sense of being connected and understood, there would be no way for you to process your emotional distress. If this were your history, after you became a parent yourself separation experiences might be an issue for you that could evoke a range of emotional responses. They might trigger your own sense of abandonment and make you feel uneasy about leaving your child. Such a discomfort would be perceived by your own child and create a sense of insecurity in him, increasing his distress, and then heightening further your own sense of discomfort in leaving him. In this manner, a cascade of emotions would be released in a chain reaction of responses that reflect your own early history. Of course, without reflection and your own self-understanding, this cascade of reactions would just be experienced in the here and now as "normal" aspects of the difficulty of separation. Self-understanding can pave the way to resolve these leftover issues.

Leftover issues often affect our parenting and cause us and our children needless frustration and conflict. Here's an issue from Mary's experience as both a mother and a child.

## SHOPPING FOR SHOES

As a parent, I discovered various leftover issues from my own childhood that were affecting my relationship with my children and robbing us of what could have been enjoyable experiences. Shopping for shoes was one of them. I found that I dreaded seeing their tennis shoes wearing out because it meant that I would have to take my two sons to the shoe store. They loved getting new shoes and initially looked forward to this excursion with excitement, as most children do. This had the possibility to be a delightful outing, since choosing new shoes is something children usually like to do, but it never turned out that way.

My sons would choose the shoes they wanted, which I verbally encouraged them to do. Even though they were quite enthusiastic about their choice, I began to spoil the experience with my doubts about the color, the price, the size of the shoes, or whatever tangible aspect I could wrap my mind around. Their excitement about their choice began to fade and an accommodating attitude of "Whatever *you* want, Mom, is okay with me" took its place. I'd vacillate and reconsider the advantages of one pair of shoes over the other and after a great deal of hassle we'd leave the store with our purchases. We were all exhausted. The excitement of getting new shoes became buried in the unpleasant memories of the experience.

I didn't want to act the way I did but repeated the experience many times, often apologizing to my children as we left the store. I always ended up in emotional conflict. "Over shoes," I'd reprimand myself.

"How ridiculous." Why did I keep repeating a pattern that I clearly wanted to change?

One day, after another disappointing shopping trip, my six-year-old son, obviously feeling deflated, asked, "Didn't you like to get new shoes as a kid?" An overwhelming "No" flooded through my body as I remembered the frustration-filled days of my own childhood shoe-shopping outings.

I was one of nine children. With so many shoes to buy, my mother always went when there was a sale, preferably a big one, and the stores were crowded with shoppers and the prices were to her liking. I never went shopping alone with my mother, since there were always three or four of us who were in need of shoes at the same time. So at a crowded sale, with mixed emotions I'd seek my next pair of shoes. I knew that I was unlikely to get what I wanted. I was unlucky enough to have a perfectly average-sized foot, for which the pickings were always very slim by sale time. My choices were minimal and I usually fell in love with something new that wasn't on sale. This choice was sure to be nixed by my mother.

Then there was my older sister, who had a very "special" narrow foot and was always allowed to purchase whatever she wanted because her size was seldom on sale anyway. I felt angry and neglected but I was told to be grateful that I was easy to fit. By the time my mother had fitted all of us, she was extremely exhausted and very irritated. Her indecision about making choices, as well as her anxiety about spending money, became full blown and I worried about her behavior. I was lost in a sea of emotions and just wanted to go home and avoid the whole shopping scene. What could have been an adventure in choosing something of my very own was spoiled.

Here I was, years later, with a mental model of shoe shopping that was bringing to my own children the same anxiety I had felt as a child. My mother was too busy getting all of us, along with our purchases,

PARENTING FROM THE INSIDE OUT

to the car to listen to or even notice my distress at the shoe shop. Because it was brought to my conscious attention by my son's question, I was able to recall my own early experiences and anxiety, which were now affecting my behavior with my children and keeping me from making this an enjoyable experience. It wasn't the shoe-shopping experience I was having in the present but the many that I'd had in the past that were influencing my behavior. I was responding to leftover issues.

Unresolved issues are similar to leftover issues, but they are more extreme, involving a more disorganizing influence on both our internal lives and our interpersonal relationships. Experiences that were profoundly overwhelming and may have involved a deep sense of helplessness, despair, loss, terror, and perhaps betrayal are often at the root of unresolved conditions. As an example, we can again use the issue of separation, but this time under a more extreme circumstance. If a young child's mother is hospitalized for an extended period of time for depression and the child is shuffled from one caregiver to another, the child will experience a deep sense of loss and despair. Separation may continue to create anxiety and affect the child's ability to have a healthy separation from her own child later in life. As a parent she may also have difficulty connecting to her child, since her own attachment was broken abruptly and she received no support. When the child becomes a parent without the opportunity to process these events and make sense of these frightening early experiences, emotional, behavioral, perceptual, and bodily memories may continue to intrude on her life. These unresolved issues can profoundly impair the parent-child relationship.

As parents we are especially vulnerable to responding on the basis of our past issues during times of stress. Here is a story of an unresolved issue that Dan became aware of shortly after he became a father.

I used to feel a strange sensation when my son was an infant and he would be inconsolable in his crying. I was surprised at the panic that would come over me as I became filled with a sense of dread and terror. Instead of being a calm center of patience and insight, I became fearful and impatient.

I tried to look within myself to understand these sensations. I thought about the possibility that I might have been allowed to cry for long periods in my own infancy. I couldn't recall this directly, but knew that the normal process of early childhood amnesia would prevent me from having a consciously accessible autobiographical recollection of such an early experience. I couldn't come up with any other plausible interpretation for this panic.

I tried it on as a narrative: "Yes, I must have been terrified of my own crying as a child. I must have had to adapt to the feeling that I was being abandoned. When my son cries now, it reactivates my emotions of fear and I experience the associated panic." I thought about it long and hard. I had no feelings about the accuracy of the story. No images. No sensations. No emotions. No behavioral impulses. In other words, this narrative elicited no nonverbal memories. The explanation also did absolutely nothing to change the panic. I felt that this did not mean that it was necessarily untrue—just that it was not helpful at that point in my journey to understand.

I was with my infant son one day when he began to cry. I felt helpless to console him and I began to have that strange panicky feeling of needing to flee. Then an image came to my mind first as a sensation of fullness in my head. The panic began to feel centered, less widespread. Then I began to see something internally that competed with what I was seeing externally. I say internal and external now, but then they felt similar: like seeing videotape with a double exposure. I closed my eyes. The external view disappeared and the internal one became clear.

I saw a child on an examining table, screaming, with a look of terror on his scrunched-up reddened face. My pediatrics internship partner was holding down his body. I had to not hear the child's screams. I had to not see his face. I could see the room. It was the treatment room of the pediatrics ward of the hospital where we had to take the kids who needed their blood drawn. It was the middle of the night and we were on call and had been awakened to figure out why this little boy had a fever. He was burning up and we had to draw his blood to rule out an infection.

The kids at the UCLA Medical Center, as in any teaching hospital, were very ill. Many of them were veterans of hospital life, but that didn't lessen their fear—the frequent blood draws just perpetuated it, and in the process destroyed their veins. My partner and I had to draw blood every night that we were on call. Now it was my turn to draw.

When a child's arm veins are so scarred that you can't draw blood from them, you have to find another vein. Sometimes it took many tries at different locations. We would trade off between who was holding the syringe and who was holding the child. We had to shut off our ears, and harden our hearts. We had to look away from the fear on the child's face, not feel the tears rolling down over our hands, and not hear the cries echoing in our ears.

But now I could hear the screams. No blood came. I had to find another site. "Just one more time," I'd tell the child, who couldn't hear me or if he could hear, he couldn't understand. He was feverish and sick, frightened, thrashing, screaming and inconsolable.

I opened my eyes. I was sweating. My hands were trembling. My six-month-old son was still crying. And so was I.

I was shocked by the flashback. I hadn't thought much of that pediatrics internship years ago except that it was a "good year," and I was glad when it was over. Over the days that followed the flashback, I thought a lot about those images. I spoke with a few close friends and colleagues about my experience. When I would begin to talk about the on-call nights, I'd feel a queasy sensation in my stomach. My hands would ache and I'd feel as if I was getting the flu. As the images came out, I would feel desperate and frightened and flooded with the scene of those young kids. I would sink into the memory: "I can't look at the child, I have to get the blood sample." I'd try to avert my gaze, in both the memory and in talking with my friends. I felt ashamed and guilty for inflicting pain. I remembered having a sensation of panic I'd have to squelch when the beeper went off at night. There was no time to talk about how much pain these children were in or how frightened of us they were. There were no opportunities to reflect on how over-whelmed and scared we were. We had to keep on going; pausing to reflect would have made it too painful to continue.

Why didn't this "trauma" from years earlier surface as a flashback, emotion, behavior, or sensation before the birth of my son? This question raises issues about memory retrieval and the unique configuration of unresolved traumatic memory. Several factors make the retrieval of a certain memory more likely. These include the associations linked to the memory, the theme or gist of the experience, the phase of life of the person who is remembering, and the interpersonal context and the individual's state of mind at the time of encoding and at recall.

I am the youngest in my family and I had no young children in my life before the birth of my son, and therefore I was never in the presence of an inconsolably crying child after my pediatrics internship. Once I found myself with a persistently crying child, I began to have an emotional response of panic. The panic can be seen as a nonverbal emotional memory activated by the context of being with a crying child. Once the panic came, my mind's recollection process initially searched for an autobiographical memory and found nothing. At the time, there was no thematic narrative memory in which the pediatrics year could have been woven. The year was "fun and over" and I did not consciously reflect on it. Then the flashback occurred.

There is often a reason why traumatic experiences are not processed in a way that makes them readily available for later retrieval. During the trauma, an adaptation to survive can include the focusing of attention away from the horrifying aspects of an experience. Also, it may be that excessive stress and hormonal secretion during a trauma directly impair the functioning of parts of the brain necessary for autobiographical memories to be stored. After the trauma, recollection of those details encoded in only nonverbal form will likely evoke distressful emotions that can be deeply disturbing.

My empathic connection with the children's terror in the hospital was overwhelming. The year was so intense, the work so demanding, the number of patients so high, the turnover so quick, and the illnesses so severe that my coping skills were put on high alert. I had distressing feelings of shame and guilt about having been the source of the children's pain and fear. Once my internship was done, I suppose I could have said, "Okay, let me now try to remember all the pain I had to cause those very sick children." Instead, I didn't reflect on that year of pediatrics and I moved on to study trauma.

As interns we attempted to avoid the overwhelming awareness of the patients' passive, helpless, and vulnerable experience by identifying ourselves only as active, empowered, and invulnerable medical workers. The child's vulnerability became a threat to our active but nonconscious effort to avoid our feelings of vulnerability and helplessness. In retrospect, the children's vulnerability became the enemy. There was often little we could do to cure their devastating illnesses, and our inability to help them added to the overwhelming sense of sadness and despair that we felt.

We were fighting against disease, fighting the existential reality of death and despair during that relentless and sleepless year. Helplessness had to become the furthest thing from our conscious minds or we would have collapsed. Vulnerability became the target of our anger toward the villain of disease we could not conquer.

This unresolved issue presented itself to me as a vulnerable first-time parent, and I had intense and shameful emotional responses to my son's crying and his vulnerability—almost finding it intolerable—and to my sense of helplessness to soothe him. Fortunately, through painful self-reflection, I was able to see this as an unresolved issue in myself and not as a deficit in my son. And this understanding allows me to easily imagine how having an emotional intolerance for helplessness can lead to parental behaviors that target that helplessness in children and attack them for it. Even with love and the best of intentions, we may be filled with old defenses that make our children's experiences intolerable to us. This may be the origin of "parental ambivalence." When their lives provoke the intolerable emotion in us, our inability to be aware of it consciously and to make sense of it in our own lives leaves us at risk of being unable to tolerate it in our children. This intolerance can take the form of becoming blind to or ignoring our children's emotions, which gives them a sense of unreality and disconnects them from their own feelings. Or our intolerance may lead to a more assertive act such as irritability or an outright though not consciously intended attack on the child's emotional state of vulnerability and helplessness. Then, the unsuspecting child becomes the recipient of hostile responses that become woven into his internal sense of identity and directly impair his ability to tolerate those very same emotions in himself.

If we have leftover or unresolved issues, it is crucial that we take the time to pause and reflect on our emotional responses to our children. By understanding ourselves we give our children the chance to develop their own sense of vitality and the freedom to experience their own emotional worlds without restrictions and fear.

## FORMS OF MEMORY

Why do we have unresolved and leftover issues? Why do events from the past influence the present? How does experience actually have an impact on our minds? Why do past events continue to influence our present perceptions and shape how we construct the future?

The study of memory provides exciting answers to these fundamental questions. From the beginning of life, our brains are able to respond to experience by altering the connections among neurons, the basic building blocks of the brain. These connections constitute the structure of the brain, and are believed to be a powerful way in which the brain comes to remember experience. Brain structure shapes brain function. In turn, brain function creates the mind. Although genetic information also determines fundamental aspects of brain anatomy, our experiences are what create the unique connections and mold the basic structure of each individual's brain. In this manner, our experiences directly shape the structure of the brain and thus create the mind that defines who we are.

Memory is the way the brain responds to experience and creates new brain connections. The two major ways connections are made are the two forms of memory: implicit and explicit. Implicit memory results in the creation of the particular circuits of the brain that are responsible for generating emotions, behavioral responses, perception, and probably the encoding of bodily sensations. Implicit memory is a form of early nonverbal memory that is present at birth and continues throughout the life span. Another important aspect of implicit memory is something called mental models. Through mental models our minds create generalizations of repeated experiences. For example, if a baby feels consoled and comforted when her mother responds to her distress, she will generalize the experience so that the presence of her mother gives her a sense of well-being and security. When distressed in the future, her mental model of her relationship with her mother will become activated and lead her to seek her out in order to be calmed. Our attachment relationships affect how we see others and how we see ourselves. Through repeated experiences with our attachment figures, our mind creates models that affect our view of both others and ourselves. In the example above, the child views her mother as safe and responsive and views herself as capable of impacting her environment and getting her needs met. These models create a filter that patterns the way we channel our perceptions and construct our responses to the world. Through these filtering models we develop characteristic ways of seeing and being.

The fascinating feature of implicit memory is that when it is retrieved it lacks an internal sensation that something is being "recalled" and the individual is not even aware that this internal experience is being generated from something from the past. Thus, emotions, behaviors, bodily sensations, perceptual interpretations, and the bias of particular nonconscious mental models may influence our present experience (both perception and behavior) without our having any realization that we are being shaped by the past. What is particularly amazing is that our brains can encode implicit memory without the route of conscious attention. This means that we can encode elements into implicit memory without ever needing to consciously attend to them.

After the first birthday the development of a part of the brain called the hippocampus establishes a new set of circuitry that makes possible the beginning of the second major form of memory, explicit

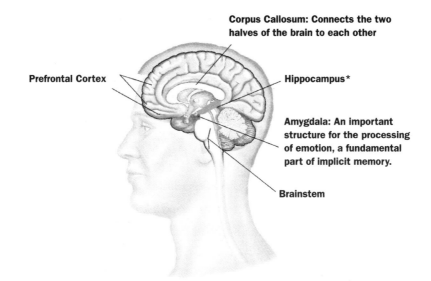

Corpus Callosum: Connects the two halves of the brain to each other

Prefrontal Cortex

Hippocampus*

Amygdala: An important structure for the processing of emotion, a fundamental part of implicit memory.

Brainstem

FIGURE 1. Diagram of the human brain looking from the middle toward the right side. Some key structures involved in memory are noted, including the Amygdala (implicit emotional memory processing), the Hippocampus (explicit forms of memory), and the Prefrontal Cortex (explicit autobiographical memory). Coherent life stories, as described in the next chapter, may involve the integration of information across the hemispheres via the Corpus Callosum.

*The shaded area represents where the hippocampus would be located on the other side of the brainstem in this diagram. At the head of the hippocampus is the emotion processing amygdala. Both of these structures are a part of the medial temporal lobe that lies just to the sides of this midline view.

memory. There are two components of explicit memory: semantic, or factual, memory, which becomes available at around a year and a half of age, and autobiographical memory, which begins to develop sometime after the second birthday. The period before autobiographical memory is available is called childhood amnesia and is a developmentally universal phenomenon occurring across cultures; it has nothing to do with trauma but instead appears to be dependent on the fact that maturation of particular structures in the brain has not commenced. In contrast to implicit memory, when explicit memory is recalled it does have the internal sensation of recollection. For both forms of explicit memory, conscious attention is required for the encoding process.

The unique feature of autobiographical memory is that it involves a sense of self and time. Autobiographical memory requires a part of the brain to mature sufficiently, around the second birthday, to allow this form of recollection to occur. This part of the brain is called the prefrontal cortex because it is at the very front of the front part of the highest layer of the brain, the neocortex. The prefrontal cortex is extremely important for a wide range of processes, including autobiographical memory, self-awareness, response flexibility, mindsight, and

## TABLE I. FORMS OF MEMORY

IMPLICIT MEMORY

- Present at birth
- No sense of recollection present when memories recalled
- Includes behavioral, emotional, perceptual, and possibly bodily memory
- Includes mental models
- Conscious attention is *not* required for encoding
- Does *not* involve the hippocampus

EXPLICIT MEMORY

- Develops during the second year of life and beyond
- Sense of recollection present when recalled
- If autobiographical, a sense of self and time are present
- Includes semantic (factual) and episodic (autobiographical) memory
- Requires conscious attention
- Involves the hippocampus
- If autobiographical, also involves the prefrontal cortex

the regulation of emotions. These are the very processes that are shaped by attachment. The development of the prefrontal cortex appears to be profoundly influenced by interpersonal experiences. This is why our early relationships have such a significant impact on our lives. However, this important integrating part of the brain may also continue to develop throughout the life span, so we continue to have the possibility for growth and change.

### FINDING RESOLUTION

Now that we have some understanding of how the brain encodes memory we can look at how to bring resolution to unresolved issues. In Dan's case, creating a plausible story about childhood amnesia had no emotional impact and was useless to alter his experience. There may have been experiences during that early period of life that were inaccessible to him explicitly but may have implicitly shaped the emotional intensity of the internship year. Without continued reflection the panic and irritation could have continued to dominate Dan's parenting style, keeping him ineffective at soothing his child's distress. He might have nonconsciously come to feel threatened by vulnerability and helplessness. Such an internal, nonreflected emotional process could have become an organizing theme in his relationships, leading Dan to discourage his son's normal dependency and push him toward premature autonomous functioning. Dan's rationalization of these experiences would have generated an attitude whereby "unresponsive children who cry too much" are seen to be spoiled and "needy." Without reflection, he would have ignored his own unresolved issue and continued to be irritated with his son.

Parental ambivalence comes in many forms, often derived from unresolved issues. Parents can find themselves filled with conflicted feelings that compromise their ability to be open and loving to their child. With defenses rigidly constructed in our own childhood and beyond, we can become frozen in our ability to adapt to the new role of caring for our children in a consistent and clear manner. Normal aspects of our children's experience such as their emotionality, their helplessness and vulnerability, and their dependence on us can feel threatening and become intolerable.

Dan continues his story: While I wanted to be connected to my son during the moments of distress, my own ambivalence created conflict between my desired response and actual behavior. Instead of being receptive and soothing, I was the source of impatience and irritability. Once I had a conscious awareness of this issue I could do something about it.

I talked to my friends about my flashback and the recollections of my internship year. I wrote in my personal journal, aware that research had demonstrated that the writing down of material about emotionally traumatic experiences can lead to profound psychological and physiological changes associated with resolution. What came out in the talks and walks and writing was how frightening and raw it all was. I had visceral responses. I felt sick. My arms would tremble and my hands would ache.

At first when my son cried I continued to feel panicky and irritated. I would say to myself, "This is a feeling from my internship and is not about my son." The panic would remain but I felt a bit better somehow. As the days wore on and the talking and writing about the internship continued, I could sense the importance of recognizing, accepting, and respecting vulnerability and helplessness—in my son and in myself. The panic and irritation became markedly less. I needed to remind myself that I was not the cause of his crying, and that being vulnerable and needy was a normal part of being a child. Making sense of my past freed me to accept my infant son's crying as well as my own feelings of vulnerability in learning to soothe him and to become his father.

The flashback has never returned. The disabling distress of that panic has not come back. The configuration of an implicit-only memory is now processed also in explicit form. The change occurred by means of the conscious processing of the implicit memories as they became integrated into a larger, explicit autobiographical narrative of that year. My life story had to embrace the emotional issues about vulnerability and helplessness at the heart of that experience in order to bring resolution.

## MOVING ON

When parents don't take responsibility for their own unfinished business, they miss an opportunity not only to become better parents but also to continue their own development. People who remain in the dark about the origins of their behaviors and intense emotional responses are unaware of their unresolved issues and the parental ambivalence they create.

Life is full of times when we must adapt quickly and do the best we can in difficult situations. Most of us have leftover or unresolved issues ready to challenge us on a regular basis. An unresolved issue can make us quite inflexible with our children and often unable to choose responses that would be helpful to their development. We're not really listening to our children because our own internal experiences are being so noisy that it's all we can hear. We are out of relationship with them and we will probably continue taking the same actions that are unsuccessful and unsatisfying to us and to our children because we're stuck in reactive responses based on our past experiences.

When we get overwhelmed and become immersed in our implicit memories of painful events or unresolved losses it is difficult to be present with our children. Our automatic adaptations to these earlier experiences then become "who we are" and our life story becomes written for us, not by us. The intrusions of unresolved issues can directly influence how we know ourselves and how we interact with our children. When unresolved issues are writing our life story, we are not our own autobiographers; we are merely recorders of how the past continues, often without our awareness, to intrude upon our present experience and shape our future directions. We are no longer making thoughtful choices about how we want to parent our children, but rather are reacting on the basis of experiences in our past. It's as if we forfeit our ability to choose our direction and put ourselves on automatic pilot without even knowing where the pilot is taking us. We often try to control our children's feelings and behavior when actually it is our own internal experience that is triggering our upset feelings about their behavior.

If we pay attention to our own internal experiences when we are feeling upset by our children's behavior we can begin to learn how our actions interfere with the loving relationship we want to have with our children. With resolution of our own issues comes greater choice and flexibility in how we respond to our children. We can begin to integrate memories into our life stories that make sense of our experiences and support healthy development for our children and for ourselves.

## INSIDE-OUT EXERCISES

1. Write in your journal when your emotions are reactive and heating up. You may notice certain patterns of interaction with your child that trigger these emotional experiences. For now, just notice them—don't try to change your response yet, just observe.

2. Expand your observations to include reflections on the possible implicit nature of your reactions to your child. Reflect on the implicit elements of memory and on the fact that these do not have the sensation of being "recollections." Making the implicit explicit, focusing conscious attention on these more automatic elements from the past, is an important part of deepening your understanding of yourself and your ability to connect with your child.

3. Think of an issue in your life that is impairing your ability to connect flexibly with your child. Focus on the past, present, and future aspects of this issue. Do any themes or general patterns come to mind from past interactions? What implicit emotions and bodily sensations emerge when this issue comes to your mind in the present? Are there other times when you have experienced these feelings? Are there elements of your past that may contribute to them? How do these themes and emotions influence your sense of your self and your connections with your child? How do they shape your anticipation of the future?

**Science and Knowledge**

People have been interested in understanding the world for as long as we have recorded history. As scientific technology has advanced, the kinds of questions people have been able to ask and attempt to answer have gotten more sophisticated, the tools available more complex and technical, and the domains of exploring knowledge more diverse. There are thousands of professional journals and scores of subspecialties within dozens of academic fields of science that actively seek to understand the world in which we live.

In this book, we will be using an interdisciplinary approach to gaining knowledge, as explored in Dan Siegel's *The Developing Mind,* which is based on the belief that there is a "reality" about the world, including human experience, that with careful study can be known in depth. However, any single approach to understanding is limited; like in the old Indian fable in which the blind men are feeling their way around parts of the elephant, any one experience or perspective can only reveal part of the larger reality. When each blind man contributes his experience of the elephant, a picture of the whole elephant begins to emerge.

An interdisciplinary view is aimed at finding the convergence among independent fields to enable a unity of knowledge to emerge. The evolutionary biologist E. O. Wilson, in his book *Consilience,* has written that this unity of knowledge is not easily achieved in academic settings, given the separations between scientific disciplines. However, an interdisciplinary approach bridges these separations and enables science to progress.

Each discipline of research, each source of knowledge, has its own particular approach, concepts, vocabulary, and ways of asking questions. An interdisciplinary approach gives equal respect to all contributors, acknowledging that interdisciplinary cooperation is the way to deepen our view of the larger reality that we seek to understand. Such an approach requires that we be humble and open, seeking to reach across disciplines in an effort to know what the elephant is really like.

The sciences we will draw on range from anthropology to psychology, brain science to psychiatry, linguistics and education to the study of communications and complex systems. One institution where this collaborative approach is followed is the Mindsight Institute. Here, training and education are provided for those interested in such an interdisciplinary process, including parents, educators, clinicians, and the general public, in hopes of nurturing a new approach to promoting well-being in individuals, families, communities, and organizations.

### Attachment, Mind, and Brain: Interpersonal Neurobiology

For many millennia, people have tried to understand the essence of being human. The human psyche, defined as the soul, the intellect, and the mind, is a functioning entity that is believed to be a process that emerges in part from the activity of the brain. The brain, an integrated system of the body itself, has been explored in the exploding new findings of neuroscience. Brain science explores the ways mental processes are created by the activity of neurons firing in the brain.

At the same time, the independent field of psychology has been exploring matters human in various dimensions: memory, thinking, emotions, and development, to name just a few. Our understanding of child development has been greatly expanded by a branch of the field of child development, attachment theory. Research into attachment has provided new understanding of how the ways parents interact with their children influence the children's later developmental pathways. Interpersonal relationships and the patterns of communication that children experience with their caregivers have been shown to directly influence the development of mental processes.

We can thus juxtapose knowledge of how the brain gives rise to mental processes (neuroscience) with knowledge of how relationships shape mental processes (attachment research). This convergence is the essence of our scientific approach, called "interpersonal neurobiology," and it provides a framework from which we can understand the day-to-day experience of children and their parents.

An interpersonal neurobiology approach to development is based on the following basic principles:

- The mind is a process involving the flow and regulation of energy and information.

- This flow of the mind occurs both in our nervous systems and in communication within our relationships—the mind is both embodied within us and embedded between us.

- The mind develops as the genetically programmed maturation of the brain responds to ongoing experience, including the experiences we have in our social relationships and our own reflective processes.

Though scientists believe that the pattern of firing of the neuronal networks is what gives rise to the "mind"—yielding processes such as attention, emotion, and memory—we don't know exactly how brain activity produces the subjective experience of mind. Also, since our minds are shaped by our social communication, the mind can be seen as a relational process as much as it is one dependent upon our body's nervous system, including the brain. One way of translating between brain and mind is to consider the mind as a process involving the flow of energy and information. For example, energy that you can observe in the mind would be the physical property of the volume of your voice, the state of alertness or sleepiness you might be experiencing, or the intensity of your level of communication with another person. A neuroscientist would examine how much energy is being utilized in various parts of the brain, using a brain scan (showing where various chemicals are consumed or blood flow is increased, as determined by increased metabolism in particular regions) or an electroencephalogram, or EEG (revealing electrical brain-wave patterns). The flow of information in the mind would be the meaning of these words you are reading—the meaning, not the ink on the page or the sound of the words. As Mark Twain suggested regarding getting the words just right, there is a big difference between lightning and lightning bug. Meaning is a powerful aspect of information processing in the mind. How we symbolize the world directly influences how we construct our perception of reality. For the brain, information is created by patterns of neuronal firing in various circuits. The location determines the kind of information (vision versus hearing); the particular pattern determines the specific information (visualizing the Eiffel Tower, not the Golden Gate Bridge).

At birth, humans are some of the most immature of beings; human infants are born with brains that are very undeveloped and they depend upon adults for critical care. The development of the necessary complexity of the child's growing brain depends upon both genetic information and experiences. In other words, the immaturity of the infant's brain means that experiences will play a significant role in determining the unique features of emerging brain connections. Experience shapes even the very brain structures that will allow the perception of those experiences to be sensed and remembered.

The care that adults provide nurtures the development of essential mental tools for survival. These attachment experiences enable children to thrive and achieve a highly flexible and adaptive capacity for balancing their emotions, thinking, and empathic connections with others. Neuroscience demonstrates that these mental capacities emanate from the integration of particular circuits in the brain; independent findings from attachment research indicate what kinds of relationship experiences a child needs to thrive and to have these mental processes develop well. Putting these pieces of the blind men's elephant together, an interpersonal neurobiology approach suggests that attachment relationships are likely to promote the development of the integrative capacities of the brain in enabling the acquisition of these emotional, cognitive, and interpersonal abilities.

## Memory, Brain, and Development: How We Remember Shapes Who We Become

The science of memory is an exciting area with an explosion of new insights into the ways experience shapes the mind and the brain. We now know that experience shapes the brain throughout life by altering the connections among neurons. For a brain, "experience" means the firing of neurons as ions flow down the lengths of these long basic cells of the brain in which there are more than 2 million miles of

neuronal fibers. Each of the 100 billion neurons of the brain is connected to an average of 10,000 other neurons. This makes for an
incredibly complex spiderweb-like network of trillions of synapses, or
neuronal connections, in the brain. Some estimate the number of
firing patterns of the brain—on/off profiles of total brain activation
that are possible—to be ten times ten one million times, or ten to the
millionth power. The human brain is thought to be the most complex
thing in the universe, artificial or natural.

Science has demonstrated that the way memory works is by the
changes in connections among neurons. When neurons become active at the same time, associational linkages are made so that if a
dog bites you when you hear fireworks, your mind may come to asso-
ciate not only dogs but also fireworks with pain and fear. These linkages are made, according to the half-century-old axiom of the
Canadian physician-psychologist Donald Hebb, because "neurons
which fire together, wire together." The psychiatrist-neuroscientist
Eric Kandel won the Nobel Prize for demonstrating that when neurons
fire (are activated) repeatedly, the genetic material inside those neurons' nuclei becomes "turned on" so that new proteins are synthe-
sized which enable the creation of new neuronal synaptic connections.
Neural firing (experience) turns on the genetic machinery that allows
the brain to change its internal connections (memory).

The brain's development also comes about when the neurons
grow and create new connections with each other, so you can see
why science tells us that memory and development are overlapping
processes: experience shapes the developing structure of the brain.
Genes determine much of how neurons link up with each other, but
equally important is that experience activates genes to influence this
linkage process. It is unhelpful to pit these interdependent processes

against each other in simplistic debates such as experience versus biology, or nature versus nurture. In fact, experience shapes brain structure. Experience *is* biology. How we treat our children changes who they are and how they will develop. Their brains need our parental involvement. Nature needs nurture.

The brain is designed to generally take care of the development of the basic foundations for normal development—we just need to provide the interactive and reflective experiences, not excessive sensory bombardment or physical stimulation, which the growing social brains of children need. Parents are the active sculptors of their children's growing brains. The immature brain of the child is so sensitive to social experience that adoptive parents should in fact also be called the biological parents because the family experiences they create shape the biological structure of their child's brain. Being a birth parent is only one way parents biologically shape their children's lives.

### Experience and the Development of Memory and the Self

Memory is the way experience shapes neuronal connections so that the present and future patterns of neuronal firing in the brain are altered in particular ways. If you've never heard of the Golden Gate Bridge, then reading those words will elicit a different response in you than in someone who lives in San Francisco and can easily visualize the bridge and generate sensations, emotions, and other associations with that bridge. The two major forms of memory, implicit and explicit, are quite different. Because the infant has the neural circuitry available in a developing but already functional form for implicit memory (emotional, behavioral, perceptual, and bodily modalities), this form of memory is available from birth, and probably even before. Implicit memory also includes the way the brain creates summaries of experiences in the form of mental models.

Explicit memory utilizes basic implicit-memory encoding mechanisms but in addition processes this information through an integrative region called the hippocampus and is dependent on the maturation of that region of the brain, after the first year and a half of life. Explicit memory is not fully available until then. With the development of the hippocampus, the mind is now able to make connections between the disparate elements of implicit memory and create a contextual mapping of integrated neural representations of experience. This is the fundamental basis of factual and then autobiographical forms of explicit memory. The hippocampus thus serves as a "cognitive mapper," creating associational linkages of representations across time and across modalities of perception (sight, sound, touch) and conception (ideas, notions, theories).

By the second birthday, the further development of the prefrontal regions of the brain enables a sense of self and time to begin to develop, signaling the beginning of autobiographical memory. Before this developmental achievement, the infant is said to be in the first phase of "childhood amnesia," in which implicit memory is present but explicit autobiographical memory is not yet available. Even after the onset of autobiographical forms of explicit memory, children still have a difficult time recalling explicitly in a continuous fashion what happened to them before the age of five or so.

No one knows yet why this is true. One possibility is that the way we consolidate our memories, the way in which we integrate the vast array of stuff in our memory storage, may not mature for many of us until into our elementary school years. Explicit memory moves from initial storage in short- and then long-term memory with the activity of the hippocampus. Over time, these long-term memories become permanent via a process called cortical consolidation. One aspect of memory consolidation is that we need Rapid Eye Movement (REM) sleep in order to take memory from long-term storage into a permanent form where it becomes free of the important hippocampus for retrieval. REM sleep is when we dream. It may be that dreaming, with its integration of emotion and memory and right and left hemisphere processing, may require certain integrative circuits that have not yet matured sufficiently in the preschool child to allow for ease of accessibility to explicit autobiographical memories later on in life. Preschool children do dream and do have recall of explicit memories of lived experiences, but the proposal here is that the consolidation process may be immature at that age and not allow those long-term autobiographical memories to enter a permanent form. If an immaturity of cortical consolidation is the limiting process, then we can understand why most people beyond the preschool years have difficulty accessing earlier periods of their lives in a continuous fashion.

One way that young children process their lived experiences is through pretend play. By creating scenarios of imagined and lived experiences, they are able to practice new skills and assimilate the complex emotional understandings of the social worlds in which they live. Creating stories through play, and presumably through our dreams, may be ways in which the mind attempts to "make sense" of our experiences and consolidate this understanding into a picture of our selves in the world.

As children grow past the preschool years, the maturation of the corpus callosum and the prefrontal regions may enable a consolidation process to occur that makes sense of the self across time and creates a scaffold of self-knowledge that we call autobiographical memory. This neurobiological maturation might explain the lag in the onset of a continuous access to autobiographical recollection during that period of development. With consolidation, we likely create our autobiographical sense of self—something that is shaped by experience and continues to develop as we grow throughout life.

Stressful experiences may impact on memory differently than nontraumatic events. One way in which unresolved trauma may block the normal processing of memory is to interfere with the normal progression of memory encoding and storage. For example, an overwhelming experience might block encoding by inhibiting the hippocampal processing of an input, thus enabling implicit processing but blocking explicit processing. This can be proposed to occur by an excessive discharge of either neurotransmitters or stress hormone during the terrifying event that blocks hippocampal encoding mechanisms. Another mechanism for blockage includes the dividing of attention in which conscious awareness is focused on a nontraumatizing aspect of the environment during the experience. This situation also would allow implicit encoding while blocking the hippocampus, which needs conscious attention for explicit encoding. Either mechanism would result in the condition of implicit-only memory traces that when retrieved would flood the person's mind without a sense that any form of memory was being recalled. Furthermore, implicit memories lack associational linkages, forged by the hippocampus, which would situate them in a context of understanding. Implicit recollections without explicit processing may be the source of the experience of flashbacks in the extreme case; more commonly it may serve as the origin of rigid implicit mental models that block a parent's ability to remain flexible and attuned to a child.

## Digging Deeper

M. L. Howe, D. Cicchetti, and S. L. Toth. "Children's Basic Memory Processes, Stress, and Maltreatment." *Development and Psychopathology* 18, no. 3 (2006): 759–769.

E. R. Kandel. *Psychiatry, Psychoanalysis, and the New Biology of Mind.* Arlington, Va.: American Psychiatric Publishing, 2005.

J. LeDoux. "Rethinking the Emotional Brain." *Neuron* 73, no. 4 (2012): 653–676.

D. J. Siegel. "Memory: An Overview with Emphasis on the Developmental, Interpersonal, and Neurobiological Aspects." *Journal of the American Academy of Child and Adolescent Psychiatry* 40 (2001): 997–1011.

———. *Mindsight: The New Science of Personal Transformation.* New York: Bantam, 2010.

———. *The Developing Mind, Second Edition: How Relationships and the Brain Interact to Shape Who We Are.* New York: Guilford Press, 2012. Chapters 1 and 2.

———, founding editor. Norton Series on Interpersonal Neurobiology. A series of over three dozen books published by Norton Professional Books and written by various authors, including Louis Cozolino (on psychotherapy), Allan Schore (on affect regulation), Daniel Stern (on the present moment), Onno van der Hart (on dissociation), Connie Lillas (on infancy), Diana Fosha and Marion Solomon (on emotion), Stephen Porges (on stress and social engagement), Pat Ogden (on the body and trauma), Jaak Panksepp (on affective neuroscience), and Ed Tronick (on relationships early in life). These books for professionals explore in an in-depth manner the practical applications of this emerging new interdisciplinary approach and its application to the process of understanding human development and helping individuals grow.

E. O. Wilson. *Consilience: The Unity of Knowledge.* New York: Knopf, 1998.

# CHAPTER 2

# How We Perceive Reality: Constructing the Stories of Our Lives

## INTRODUCTION

Stories are the way we make sense of the events of our lives. Individually and collectively we tell stories in order to understand what has happened to us and to create meaning from those experiences. Storytelling is fundamental to all human cultures, and our shared stories create a connection to others that builds a sense of belonging to a particular community. The stories of a particular culture shape how its members perceive the world. In this way, stories both are created by us and shape who we are. For these reasons, stories are central to both individual and collective human experience.

We all have our individual stories, the narratives of our personal life experiences, through which we deepen our self-knowledge and develop a greater understanding of ourselves and our relationships with others. Autobiographical narratives attempt to make sense of our lives—both the things that have happened to us and the internal experiences that create the rich texture of each individual's unique, subjective sense of life. Through deepening our self-understanding, by exploring the events and mental processes of our lives, our autobiographical stories will grow and evolve.

Children try to understand and make sense out of their experiences. Telling your children the story of an experience can help them integrate both the events and the emotional content of that experience. Such an interaction with you can greatly help them make sense out of what happened to them and gives them the experiential tools to become reflective, insightful people. Without emotional understanding from a caring adult, a child can feel distress or even a sense of shame.

Annika came to Mary's nursery school at three years of age speaking only Finnish. Her family had relocated for two years while her father was a visiting lecturer at UCLA. During Annika's initial transition to school her mother stayed with her until Annika was comfortable with the environment and the teachers. She was a very sweet and outgoing child who loved to play with other children, and the language barrier was almost insignificant in their enjoyment of the activities and each other.

Her mother had been comfortably able to leave her at school for several weeks when an incident occurred that emphasized how important a story can be in helping a child to process distress. Annika had been playing happily all morning until she fell down and skinned her knee. Like most children, after an injury she cried for her mother. She was unable to take any comfort from the teacher and continued being quite distressed. The teacher asked the office assistant to try to reach Annika's mother by telephone and continued her efforts to comfort Annika. Usually retelling the story, including both the content and the emotions of the event, greatly helps a child to both understand the experiences and feel the comfort of an empathetic adult. Since the teacher didn't speak Finnish and Annika didn't understand enough English, the teacher's telling the story had little success. The teacher quickly gathered a few props and began to tell the story again with the use of several dolls and a toy telephone. A small doll represented Annika and was used by the teacher to reenact Annika's experience. The telling of a story involves the narration of a series of events and the experiences of the characters in those events. First the doll representing Annika was playing, and then she fell down. The teacher made crying sounds to symbolize Annika's crying. Annika stopped crying and watched. The story continued. The "teacher doll" talked gently to the Annika doll, and then the actual Annika started to sob again. When the "teacher doll" picked up the toy telephone to call the "mama doll," Annika again stopped crying, watched, and listened.

Using the dolls, the teacher acted out the story of the skinned knee and the call on the telephone to ask Mama to come to school to get Annika several times. Annika understood "Mama" as well as her name, and with the repetition of the story along with the visual props she began to understand what had happened and what was going to happen. Each time the story was retold Annika's distress lessened. After a brief time she got down off the teacher's lap and happily went back to her play, apparently secure that her mama would be coming to get her. When her mother arrived, Annika brought the dolls and the telephone to the teacher because she wanted to hear the story again and share the story of the experience of the skinned knee and her distress with her mother.

Telling the story gave Annika comfort and allowed her not only to make sense of what happened but to anticipate the future outcome of the arrival of her mother. As adults we usually tell our stories with words. Children, however, even those who can understand spoken language, greatly benefit from using props like dolls and puppets or drawing pictures to help them make sense of their experiences. When children understand what has happened to them and what may happen to them, their distress is usually greatly reduced.

There may be experiences from your own childhood that you

couldn't make sense of at the time, because no caring adult was available to help you understand your experience. From the beginning of life, the mind attempts to make sense of the world and to regulate its internal emotional state through the relationship of the child with the parent. Parents help children regulate their internal states and bring meaning to experience. As children grow, they develop the capacity to create an autobiographical narrative from these experiences. This ability to tell stories reflects the fundamental way that the child has come to make sense of the world and to regulate his or her emotional states.

The way we tell our life stories reveals the way we have come to understand the events of our lives. What comes up for you when you talk about the events of your life? Do you find yourself describing your experiences from a distance, or do you emotionally relive them as you tell your story? Do particular issues seem emotionally intense and unresolved, even though they may have happened long ago? Do you remember many details about your early life? What feelings come up when you tell the story of your early experiences?

Our life stories can give us clues about how our present is shaped by the past. The way we tell our stories and how we emphasize different aspects of our experiences can reveal the way we have come to understand the world and ourselves. For example, you may think of the events of your family without focusing much on the nature of the relationships among your family members. Some families have quite distant styles of relating to each other, where emotions are rarely shared and individuals live an emotionally autonomous existence. In such a family, both parent and child may have difficulty constructing a rich autobiographical story. In this situation, the details may be difficult to retrieve and the emotions sometimes absent from the story. In these families, communication is often about the external events of life rather than about the mental life of the family members. In emotionally distant families, the important capacity for mindsight, the ability to perceive one's own mind and that of others, appears to be quite minimal in both parent and child. Stories are our mind's attempt to make sense of our own and others' rich inner worlds.

## WAYS OF KNOWING

The mind, coming in part from the activities of the brain, has different modes of processing. At a basic level we have different perceptual systems, such as sight, hearing, touch, taste, and smell. On another level we have the various forms of "intelligence," including linguistic, spatial, kinesthetic, musical, mathematical, intrapersonal, and interpersonal. The mind is complex and reveals itself in a myriad of wondrous and distinct ways of perceiving and interacting with the world. How we perceive directly influences how we behave. As organisms with input and output pathways, we have brains that are designed to take in data from the world, process them internally (often called cognition), and then create a specific response. *Input–internal processing–output.* This is the most basic way of describing the role of the brain and overall nervous system.

We can examine how the left side of the brain differs from the right in its internal processing of incoming data. The right and left sides of the brain evolved over millions of years from the asymmetrical nervous systems of lower animals and are quite distinct from each other. The two physically separated sides of the brain are connected to each other through bands of neural tissue called the corpus callosum. This separation enables each side of the brain to function somewhat independently and to carry out quite distinct forms of processing. With the passage of neural information back and forth between the two hemispheres, an integrated form of processing is also possible that enables the brain to achieve higher levels of functioning. Each side of the brain perceives uniquely and processes information differently. The benefit of these differences is that we get more functions out of a single brain when its parts are specialized in their functioning. When each differentiated side can contribute to an integrated whole, we are capable of achieving

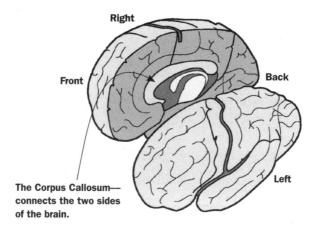

FIGURE 2. Side View of the Divided Brain

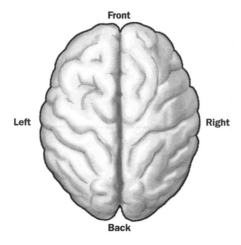

FIGURE 3. Top View of the Brain

## TABLE 2. RIGHT AND LEFT MODES OF PROCESSING

RIGHT-MODE PROCESSING

Nonlinear

Holistic

Contextual

Specializes in

> Autobiographical information
>
> Sending and perceiving of nonverbal signals
>
> Spontaneous and raw emotions
>
> Awareness, regulation, and integrated map of the body
>
> Social cognition and mindsight—understanding others

May involve a predominance of the brain's right hemisphere in processing

LEFT-MODE PROCESSING

Linear

Logical

Linguistic

Specializes in

> Syllogistic reasoning—looking for cause-effect patterns
>
> Linguistic analysis—using words to define the world
>
> "Right-versus-wrong" thinking

May involve a predominance of the brain's left hemisphere in processing more than the individual sides in isolation could achieve. If both sides of the brain were the same, we'd be less complex and adaptive.

One way of referring to these two distinct aspects of perception—internal processing—output is with the terms right and left *modes* of processing. The right mode, primarily created by the activity of the right hemisphere of the brain, has a nonlinear, holistic manner of processing information. This mode is specialized for taking in and processing contextual and spatial information. Autobiographical data, the processing and sending of nonverbal signals, an integrated sense of the body, mental models of the self, spontaneous emotions, and social understanding are all processed predominantly on the right side.

In contrast, a left mode of processing, taking place primarily on the left side of the brain, involves something quite different: linear, logical, language-based processing. "Linear" means one piece of data follows another in a line. "Logical" signifies a search for cause-effect relationships as patterns in the world. Language-based processing utilizes the digital (yes/no, on/off) bits of information contained within words, such as these you are reading on this page.

## NARRATIVES AND THE BLENDING OF LEFT AND RIGHT MODES

In *The Developing Mind,* a proposal is made that narratives that make sense of life emerge out of a blending of the left-mode drive to explain and the right-mode storage of autobiographical, social, and emotional information. A coherent narrative, one that makes sense of life experiences, may emerge from a flexible blending of the left and right modes of processing. When the left-mode drive to explain and the right-mode nonverbal and autobiographical processing are freely integrated, a coherent narrative emerges.

As a parent, making sense of your life is important because it supports your ability to provide emotionally connecting and flexible relationships with your children. Having a coherent sense of your own life history enables you to offer the kind of experiences that help children make sense of their own lives. Making sense of your life can be enhanced by learning about what the left and the right mode of processing feel like for you. Each of us has the ability to distance ourselves from the uncontrollable, less predictable free-form sensations of our raw emotional experiences. This distancing can be seen as a predominantly left-mode processing state: the left mode is overriding input from the right at that moment.

In contrast, there may be other times when we may become aware of being filled with sensations that we cannot name or understand logically. There may be images in the mind's eye, bodily sensations, or notions that float in and out of our consciousness. We may lose track of time and feel closely connected to what we are perceiving, having little concern for the cause-effect relationships of the world around us. We may become immersed in the bodily, emotional, and perceptual aspects of an experience, which all are aspects of right-mode processing.

When we attempt to weave a story about our lives with left-mode processing, we use its language-based, linear, logical approach to understand cause-effect relationships. To tell such a linear narrative about the self using left-mode processing we must draw on information from the right mode of processing. If such information either is not readily available or it floods in like a tidal wave, we will be unable to tell a coherent tale. Unresolved issues may create such incoherence. In such a situation, the story will be incoherent either because it lacks emotional and autobiographical richness and meaning, or because it cannot make sense of these right-mode contributions to the story. Stories that make sense of our lives require both clear thinking and access to the equally important emotional and autobiographical aspects of experience.

For example, a woman in her midthirties whose father died when she was an adolescent had never fully grieved her loss. When she talked about her teen years, she would suddenly become flooded with tears and be unable to continue telling her life story. Her last experience with her father had been filled with tension as they fought about his disapproval of her boyfriend just before his fatal heart attack. Not until she could fully deal with her sense of guilt could she fully mourn his death and integrate this loss into a coherent narrative of her life.

As you continue the process of self-reflection, try to be aware of what may feel like quite distinct modes of subjective experience. Blending these two modes together may be essential for the creation of an integrated coherence of mind, and the telling of a coherent life story.

## COHERENCE AND INTEGRATION

Our relationship with our children is built upon our many shared experiences. Experiences that draw on a wide range of internal processes appear to promote well-balanced interpersonal interactions. When separate and distinct "differentiated" processes are brought into a functioning whole, we can use the term "integration." Let's consider the possible ways that the brain can become integrated. When the more complex, reflective, and conceptual processing of the anatomically higher cerebral cortex is combined with the more basic, emotional, and motivational drives of our deeper brain regions, we achieve the ability to respond from an integrated state, a vertically integrated, higher mode of processing. When the reflective functions of the cortex are cut off, we become inflexible as we enter the nonintegrated "lower mode" of response.

There can also be horizontal integration in which the left and right sides of the brain work together. This bilateral integration may be at the core of how we create the coherent narratives that emerge when we make sense of our lives. As coherent narratives are the best predictor of a child's having a secure attachment to us, this bilateral integrative process may be at the heart of a parent's ability to provide a child with a nurturing environment and a secure base.

The creation of our life stories may depend upon yet another form of integration, in addition to the vertical and horizontal dimensions of integration: temporal integration, connecting processes across time. This is fundamentally what stories can do: they connect the self in the past, present, and anticipated future. This mental time travel is a central feature of stories as they are found throughout the world.

Healthy relationships, and perhaps our own internal sense of coherence and well-being, may depend upon a fluid and active integrative process within our minds. Mental well-being may depend in many ways upon the layers of integration that deepen our sense of connection to ourselves and to others. Promoting integration across these many differentiated domains may enable our self-knowledge and interpersonal relationships to flourish and enrich our lives and the lives of our children.

Stories shared between one person's mind and that of others are a universal way we connect with each other. Stories enable us to have interpersonal integration. When we think of important people in our lives, we often recall moments of connection that are expressed in the personal stories we cherish about our relationships. At weddings, graduations, reunions, and funerals, stories fill the air as people connect with each other in reflecting on the power of their shared experience as they bear witness to the passage of time.

As you reflect on your own life story, you may deepen your understanding of yourself and be able to integrate your emotions into your daily living and give respect to that valuable way of knowing. As your mind changes with self-reflection, you may find that your experiences with your child change also. Experience shapes the mind just as the mind shapes experience. The personal growth and deepening self-knowledge that come from developing your life stories can enrich your capacity for mindsight and enhance your sensitivity to your child.

**INSIDE-OUT EXERCISES**

1.  As an exercise in increasing your awareness of the different modes of processing, look over Table 2 on page 37 and write or draw in your journal examples from your own experience of each of the elements in each mode. As you deepen your self-knowledge, you may find that your sensitivity to the right and left modes of processing is heightened. Pay special attention to the nonverbal, nonlogical, often unpredictable sensations of the right mode in both yourself and your child.

2.  Writing words in a journal may pull heavily from the left mode, so please give support and respect to the images of your nonverbal right mode of processing when reflecting on your experiences. Awareness of the right mode is often more subtle than our language-based thoughts can adequately express! With these ideas in mind, notice the images and memories that emerge in trying to tell a story about the history and meaning of a particular issue affecting your connection with your child. Write about when it started, how it has shaped your development, and how it affects your present relationships with your child. What

aspects of this issue are the most challenging for you? Imagine resolving a leftover or unresolved issue. Write a new story about what this healing would be like. How will things be different now and in the future?

3. Think of three words that describe your relationship with your child. Do these words resemble those that you would use to describe memories of your childhood experiences with your own parents? How do they differ? Do these words accurately summarize these relationships? Are there parts of your memories of these important relationships that do not fit well with these generalizations? How do these exceptions fit into the broader story of your life with your parents and with your child?

 **SPOTLIGHT ON SCIENCE**

### The Science of Stories

The science of stories takes us to a wide range of academic disciplines, from anthropology and the study of culture to psychology and the study of how people relate the content of their memories to each other. Our experiences within families shape how we come to perceive the world around us. The larger culture in which we live also has profound effects on the ways our minds come to process information and create meaning in our lives. Recent findings from brain science also can contribute to our understanding of the important roles that stories play in human lives.

These sciences tell us a number of things:

- Stories are universal—they are found in every human culture on the planet.

- Stories are found throughout the human life span—they are present in interactions between adults and children early in life and continue to play a role in relationships into maturity.

- Stories may be uniquely human—no other animal appears to have the narrative instinct and storytelling drive.

- Stories involve logical sequencing of events, but they also play a powerful role in regulating emotions; in this way, stories are a good example of how emotion and analytical thinking are intertwined.

- Stories play a role in everyday communication as well as in the internal sense of self. This blend of the interpersonal and the personal, so characteristic of much of the human mind, reveals what social creatures we truly are!

- Stories may play a vital role in memory processes—a proposal that stems from the study of explicit memory and its eventual processing through dreams toward permanent, or "cortically consolidated," memory.

- Stories have been correlated with brain function. In particular, the left hemisphere appears to be driven to make guesses about the logical connection among bits and pieces of information, while the right hemisphere supplies the emotional context and autobiographical data necessary to have a personal life story make sense.

* * *

## Mental Models: How Our Experiences Shape How We See, Behave, and Create Our Own Reality

Science has shown that the brain, even in young infants, is quite capable of making generalizations, or mental models, from repeated experiences. These mental models are a part of implicit memory, and are thought to be created in the patterns of neuronal firing in the perceptual modalities of vision, hearing, touch, taste, and smell that accumulate in repeated interactions. The model that is generated in the brain serves as a kind of view, perspective, or state of mind that will directly influence the way we perceive and the way we respond in the future. Implicit memories, and especially the mental models of our many forms of lived experiences, likely create the themes of the stories we tell and organize the ways in which we make life decisions.

Mental models serve as a kind of funnel through which information is filtered, as lenses that help us to anticipate the future and therefore prepare our minds for action. These lenses are outside our conscious awareness and bias our perceptions without our having conscious knowledge of their existence. They are built from past experience, are activated in specific ways, and function quickly to shape our view of reality, our beliefs and attitudes, and the way we will engage with the world around us.

For example, if you were bitten by a cat when you were a young child, encountering one later in life may entail a rapid shift in your state of mind, an instantaneous creation of a state of fear: vigilance toward the feline crossing your path, intense focus on her teeth, an internal sense of panic, and a readying yourself to quickly respond should the cat move in your direction. These changes in your perception, your internal emotional state, and the activation of your survival responses (fight-flight-freeze-faint) are automatic. They are shaped by your brain's ability to rapidly, without your conscious attention or planning, create an organizing state of mind that shapes your perception and prepares your body for action. This is the essence of how mental models serve to funnel and filter incoming information and create a rapid shift in your state of mind.

Mental models, and the states they initiate, are a fundamental part of implicit memory. As such, we don't have a conscious sense of their origins, though we may be very aware of their effects (watching the cat, feeling fear). Issues from our past may influence us in the present and alter how we behave in the future by directly shaping how we perceive what is going on around us and inside of us. The shadows that implicit mental models cast upon our decisions and the stories we tell about our life can be made explicit through focused self-reflection. Such a conscious process can deepen self-understanding and may be a way to alter mental models and open the door to ongoing development throughout life. Self-understanding, which can liberate us from these imprisoning shadows of the past, requires that we become reflective of how these patterns of perceptual bias and behavioral impulses may be a fundamental part of ingrained mental models and inflexible states of mind. Taking time to reflect opens the door to conscious awareness, which brings with it the possibility of change.

Studies of memory encoding suggest that how you perceive is shaped by what you experience; yet perception shapes how we process what we experience. As we come to understand ourselves, both explicit knowledge and implicit memory can shape our view of who we are and organize how we interact with the world. Endel Tulving and his colleagues in Toronto have described the process of self-knowing consciousness, or *autonoesis,* that emerges from the integrative neural circuitry of the prefrontal regions of the brain. Self-understanding is created through linking of the various elements of memory stored throughout the brain, a blending of elements of the past, perceptions in the present, and anticipations of the future.

The interactive feedback loop of experience actually shaping how we perceive and how we come to anticipate the future within mental models of the world means that we are active shapers of our own construction of reality while at the same time experience shapes how we come to construct our views of that reality. Experiences that integrate our bodily experiences with those we have within relationships with important others can be a basic building block of the developing self. The experience of Helen Keller (1880–1968), blind and deaf, is described by Paul John Eakin as an example of how people narrate the stories of their lives and illustrates this developmental linkage (Eakin 1999, 66–67): The writing of Helen Keller, said Eakin, was a "rare and possibly unique account of the emergence of selfhood that occurs at the moment when language is acquired. . . . Although Keller had previously mastered a small vocabulary of finger-words spelled into her hand by her teacher, Anne Sullivan, it was only when Sullivan placed one of her hands under the spout and spelled into the other the word water that Keller achieved simultaneously a sense of language and self. It was truly a kind of intellectual and spiritual baptism: 'I knew then that "w-a-t-e-r" meant the wonderful cool something that was flowing over my hand. That living word awakened my soul.' I summarized the upshot of the well-house episode schematically as follows. . . . In its basic outline, this analysis of the Keller episode squares with the social constructivist perspective on identity formation. My schematic, freeze-frame account (self/language/other) reflects the fact that an entire passage of developmental history, normally requiring many months to transact, is compressed in Keller's case into the space of a revelatory moment. . . . Keller stresses both the relational dimension of the episode (the decisive role of 'Teacher') and the grounding of the entire experience in her body."

The prefrontal regions of the brain, especially the orbitofrontal cortex, are crucial for mediating interpersonal communication, representations of the body, and autobiographical awareness. The openness to growth of the prefrontal cortex throughout life may help us to understand how deepening self-knowledge can alter the ways we experience others and ourselves. It may be that as we develop throughout life, the experiences we have with others and within our bodies form the foundation for our emerging sense of self that grows throughout our lives.

Taking time to reflect on our interpersonal and internal experiences can enable us to grow by a deeper self-knowledge and awareness. A deeper self-understanding is also built on a coherent autonoesis—making sense of our past, our present, and how we anticipate the future. The take-home message is that we can become the active authors of our own autobiographies and thereby help our children learn to become active shapers of the ways they perceive and create their own lives!

* * *

## Logic and Story, Mind and Brain

The developmental psychologist Jerome Bruner describes two basic ways in which the mind processes information. One is a "paradigmatic" deductive mode in which a series of linearly related facts are connected by logical inferences regarding cause-and-effect relationships. This is similar to the logical, linear, language-based mode of cognition located in the left side of the brain. The other way of processing information that Bruner has suggested is the narrative mode, in which the mind processes data by creating stories. This mode, which develops earlier and is found in all cultures, is a distinct process of creating a world of possibilities, not just actualities. In Bruner's view, stories entail telling both a sequence of events and the internal mental life of the characters of the story. These narrative processes enable us to delve deeply into the subjective worlds of people.

The narrative mode of thinking does not have an easily defined locus in the brain. However, studies of related processes such as autobiographical recollection and confabulation (making up fictional accounts to appear as true ones) point to some intriguing possibilities. The left hemisphere of the brain appears to be driven to tell a story about events. This is not surprising, given that it specializes in logico-deductive reasoning that attempts to explain how things are related to each other in the world. Filled with facts, language, linear processing, and a drive to create categories, the left hemisphere has what cognitive scientist Michael Gazzaniga has called an interpreter function.

When disconnected from the right side of the brain, however, the left side makes stories up. It lacks the context to make sense of what it sees. Oddly enough, it doesn't seem to mind just stringing facts together to make them seem cohesive, though they may not fit in with the larger sense or context of a situation. In this way, stories may be cohesive (somewhat logical in their connections) but not coherent (not making sense in an overall emotionally contextual, sensory way).

How can this be?

One way to explain this finding is to look to the right hemisphere's role as a supplier of social and emotional context. The right hemisphere takes a front-and-center role in processing nonverbal signals. This side of the brain is more directly connected to the limbic circuitry of the brain which creates emotion and motivational states. For a number of reasons, the ability to sense the subjective life of another person, to perceive the person's signals and make sense out of them, seems to depend upon a functioning right hemisphere. Social, emotional, nonverbal, and contextual information serves as the raw material for "mindsight"—the capacity to perceive the mind of others and of the self.

Stories tell of sequences of events *and* the internal life of the characters in those events. Internal life is sensed and understood primarily by the right side of the brain. For a story to "make sense"— for it to incorporate the subjective/social/emotional meaning of the internal life of the characters—it must include right-hemisphere processing. For these reasons, we can suggest that for the mind to tell a coherent story, one that makes sense of one's own or others' lives, the left hemisphere must integrate itself with the right hemisphere. Thus, coherent stories likely result from bilateral, interhemispheric integration.

## Digging Deeper

A. Botzung, E. Denkova, and L. Manning. "Experiencing Past and Future Personal Events: Functional Neuroimaging Evidence on the Neural Bases of Mental Time Travel." *Brain and Cognition* 66, no. 2 (2008): 202–212.

J. S. Bruner. *Making Stories: Narrative in Law, Literature, and Life.* New York: Farrar, Straus and Giroux, 2002.

J. S. Bruner and H. Haste, eds. *Making Sense: The Child's Construction of the World.* New York: Routledge, 2011.

O. Devinsky. "Right Cerebral Hemisphere Dominance for a Sense of Corporeal and Emotional Self." *Epilepsy and Behavior* 1 (2000): 60–73.

P. J. Eakin. *How Our Lives Became Stories: Making Selves.* Ithaca, N.Y.: Cornell University Press, 1999.

P. J. Eakin. *Living Autobiographically: How We Create Identity in Narrative.* Ithaca, N.Y.: Cornell University Press, 2008.

I. McGilchrist. *The Master and His Emissary: The Divided Brain and the Making of the Western World.* New Haven, Conn.: Yale University Press, 2009.

H. J. Markowitsch and A. Staniloiu. "Memory, Autonoetic Consciousness, and the Self." *Consciousness and Cognition* 20, no. 1 (2011): 16–39.

E. Ochs and L. Capps. *Living Narrative.* Cambridge, Mass.: Harvard University Press, 2001.

D. J. Siegel. *The Developing Mind, Second Edition: How Relationships and the Brain Interact to Shape Who We Are.* New York: Guilford Press, 2012. Chapter 5.

D. J. Siegel and B. Nurcombe. "The Development of Perception, Cognition, and Memory." In Mel Lewis, ed., *Child and Adolescent Psychiatry: A Comprehensive Textbook,* third edition, pp. 228–238. Baltimore: Lippincott, 2002.

S. P. Springer and G. Deutsch. *Left Brain, Right Brain,* fifth edition. New York: Freeman, 1998.

M. A. Wheeler, D. T. Stuss, and E. Tulving. "Toward a Theory of Episodic Memory: The Frontal Lobes and Autonoetic Consciousness." *Psychological Bulletin* 121 (1997): 331–354.

# CHAPTER 3

# How We Feel: Emotion in Our Internal and Interpersonal Worlds

## INTRODUCTION

As parents we want to have loving, lasting, and meaningful relationships with our children. Understanding the role that emotions play in how we connect to each other can help us to do just that. It is through the sharing of emotions that we build connections with others. Communication that involves an awareness of our own emotions, an ability to respectfully share our emotions, and an empathic understanding of our children's emotions lays a foundation that supports the building of lifelong relationships with our children.

Emotions shape both our internal and our interpersonal experiences and they fill our minds with a sense of what has meaning. When we are aware of our emotions and are able to share them with others, our daily lives are enriched because it is through the sharing of emotions that we deepen our connections with others.

As a parent your ability to communicate about emotions supports your child in developing a sense of vitality and empathy. These qualities are important for the nurturing of close, intimate relationships throughout the life span. Nurturing relationships involve the sharing and amplification of positive emotions and the soothing and reduction of negative ones. Feelings are both the process and content of interpersonal connection between parent and child from the earliest days of life.

Imagine that your child comes in from playing in the backyard very excited about the colorful beetles he's collected in a small open glass jar. "Look, Mommy, look what I found, aren't they pretty?" All you see is the possibility of bugs loose in the house. "Get those creepy things out of here right now," you say sternly. Your child starts to protest, "But Mommy, you didn't even look, their wings are shiny green." You glance fleetingly at the jar, and taking your child by the arm you march him toward the door, reminding him that "bugs live outside and they need to stay outside."

In this situation the child's emotional experience was totally missed. His joy and delight were not shared and he probably had some confusion about the meaning and value of his experience. He felt "good" and excited about his discovery and came in to share this feeling. Instead, he was responded to as if he was "bad." A meaningful emotional connection would have given value to his experience and the parent would have shared in the child's excitement and discovery. This doesn't mean you have to live with bugs in the house, it just means that it is important to attune to, or resonate with the emotional experience inside the child before changing the external behavior. Attuning to your child's emotions can mean getting down on his level, having an open and receptive demeanor, looking at what he has brought to show you, and expressing curiosity and enthusiasm in your tone of voice. "Let me see, wow! They're colorful little beetles, aren't they? Thank you for showing them to me. Where did you find them? I think they'll be happiest living outside." Not only would this have supported the parent's relationship to the child but also, because the child would have experienced that his ideas and emotions were of value to his parent, he would have felt a stronger sense of himself. When a parent resonates with the child's emotions, the child's experience of himself is that he is "good." Emotional connections create meaning for the child and affect his understanding of both his parent and himself.

Exactly what is an emotion? We may know an emotion when we feel one, but it is very hard to explain the experience. Insight from the sciences can be quite helpful in understanding what emotions are and what a central role they play in our lives. This knowledge can be used to deepen our self-knowledge and enhance our relationships with our children and others.

Emotion is often thought of as a range of feelings that we can sense in ourselves and perceive in others, and that we can usually label with words such as sadness, anger, fear, joy, surprise, disgust, or shame. These emotions are present in people from all cultures throughout the world. However, these easily categorized emotions are only one aspect of the important role of emotions in human life.

We would like to offer an additional perspective on the whole idea of emotions. For now, leave aside the categories of emotion and open your mind to a new possibility. *Emotion can be thought of as a process that integrates distinct entities into a functional whole.* "Positive" emotional states may involve increases in integration while "negative" states may decrease integration. Joy, excitement, love, and satisfaction may each involve increases in how the many aspects of ourselves are brought together into a functional whole. That's integration. And uncomfortable or harmful states, such as prolonged anger, sadness, or fear, may constrict how the many aspects of ourselves are brought into coordination and balance. Increased integration comes along with a sense of harmony; decreased integration can lead us toward states of being rigid or of being chaotic for prolonged periods of time. That may sound pretty abstract, so let's try to bring some clarity to this viewpoint and look at its practical applications.

Emotion as a fundamental integrating process is an aspect of virtually every function of the human brain. As a collection of massive amounts of neural cells capable of firing in a chaotic fashion, the brain needs an integrating process to help it achieve some form of balance and self-regulation. Emotion is the process of integration that brings self-organization to the mind. As we discussed earlier, integration may be at the heart of a sense of well-being within ourselves and in our relationships with our children and others. How emotion is experienced and communicated may be fundamental to how we come to feel a sense of vitality and meaning in our lives.

More basic than the emotional categories is primary emotion, which can be described in the following manner. First, the brain responds to an internal or external signal with an initial orientation response that activates the mind to focus attention. This initial orientation basically says, "Pay attention now! This is important!" Next, the brain responds to that initial orientation with an appraisal of whether that signal is "good" or "bad." This appraisal is then followed by the activation of more neural circuits, which elaborate, or expand, this activation into associated brain regions. This appraisal/arousal process can be thought of as the fundamental surges of energy in the mind that accompany the processing of information. These elaborated appraisal processes are how the brain creates meaning in the mind. Emotion and a sense of meaning are created by the same neural processes. As we'll see, these same circuits of the brain also process social communication. Emotion, meaning, and social connection go hand in hand.

These initial primary emotions are the brain's first assessment of the importance and goodness/badness of an experience. Through emotion our minds become organized and prepare our bodies for action. An appraisal as good leads to approach; an appraisal as bad leads to withdrawal. In response to our questions about how they feel, children so often will respond with "Good," or "Bad," or "Okay." These simple words, frequently not accepted at face value by the parent, are actually pretty direct expressions of these primary emotional processes.

Primary emotions are directly observed in nonverbal expressions. Facial expressions, eye contact, tone of voice, gestures and touch, posture, and the timing and intensity of response reveal the surges of arousal and activation, the profiles of energy flow through the mind, that are the essence of how a person is feeling. Primary emotions are the "music of the mind."

Connecting to primary emotional states is how we tune in to each other's feelings. These primary emotions exist all of the time. We are usually more consciously aware of emotions when these primary emotions become further channeled into the more differentiated responses of the categorical emotions. We may unfortunately think of "being emotional" as only the expression of categorical emotion that can

## TABLE 3. KINDS OF EMOTION

PRIMARY EMOTION: SURGES IN THE FLOW OF ENERGY THROUGH THE MIND

Initial orientation: "Pay attention now!"

Appraisal and arousal: "Good or bad?"

CATEGORICAL EMOTION: CHARACTERISTIC MODES OF EXPRESSION SEEN IN ALL CULTURES

Discrete, basic emotions such as sadness, fear, anger, joy, surprise, disgust, and shame make us lose track of the profoundly important process of connecting with each other's primary emotional states. Relating at the level of primary emotions allows us to integrate our experiences within ourselves and with others. Attuning to each other's internal states links us in a state of emotional resonance that enables each person to "feel felt" by the other. In this resonating connection, two people mutually influence the internal state of the other. This attunement, this connecting resonance, enables us to feel joined.

## CONNECTING UNDER WATER

A father in his forties came to Dan for consultation because he was concerned that he "did not feel anything." His mother was ill and dying and his partner at work had just been diagnosed with a life-threatening illness, but he "felt nothing about it." His whole life, he insisted, was filled with moving "through space," working, doing what needed to be done, but having little that felt meaningful.

He wanted to change because he wanted a better relationship with his family and thought it wasn't right that he didn't feel anything, especially since his mother and partner were very ill. "All I do is rationalize why it will probably be okay or why it is okay even if the situation gets worse. I tell you, it just isn't right. But I just don't feel anything," he said. He often felt isolated in his day-to-day interactions with people and his life was filled with a sensation of emptiness.

He did not suffer from depression and suffered little from worries, anxieties, or any other manifestation of a psychiatric disorder for which he might be diagnosed and treated. He was an accomplished academic, had earned accolades from his colleagues, and was in high demand in his work. He was an ethical and fair person, and his associates and students seemed to do well under his leadership.

This father had to examine his earlier life experiences in order to get a framework in which to place the information about his subjective experiences from the present back to his earliest memories. He had few recollections of his experiences with his parents. What he did remember was that they both were very "intellectual" and never spoke to him about his own thoughts or feelings. They focused on accomplishments and "right and wrong behaviors" but never on the subjective, internal side of life. When his father died during his teens, he and his mother never discussed the loss.

One of the qualities that had attracted him to his wife was her emotional intensity. The oldest of his three children was now entering adolescence and he wanted to feel closer to her as well as his two younger children. When asked how he felt about them, his eyes became watery as he talked about the birth of his first child. He said that he had experienced the most intense feeling he has ever had: he felt so in love with his daughter, and yet so scared of how responsible he was for her welfare, that he could hardly tolerate the feelings.

He had been quite artistic as a child and was very interested in the aesthetic value of things—how they appeared, how objects were positioned, how colors went together. He had thought of becoming an architect but chose an academic research career instead.

An explanation that this man's right hemisphere was quite underdeveloped didn't fit with his well-developed artistic sensibility. However, his lack of empathic connection, the impairment in his utilization of his capacity for "mindsight," and his general sense of disconnection from his own internal world were suggestive of impairment in the integration of his right and left modes of processing.

In therapy he was encouraged to write in a journal, since self-reflective writing inherently requires the collaboration of the right and left hemispheres. In his weekly sessions, in order to further the attempt to make a connection between his right and left hemispheres, the focus was on his bodily sensations and his visual imagery and on exploring what these nonverbal experiences may mean.

After he had been working in therapy for several months he had a new experience in relating to his daughter. While on a trip to Hawaii, the two of them had gone snorkeling and as they were looking for fish in the coral reef, they swam together and signaled to each other using gestures and eye contact through their masks. The man described having an overwhelming feeling of connection to his daughter during this experience and he was filled with a sense of joy in their shared exploration of the underwater world. On reflection, he said, "I'll bet it had to do with our nonverbal communication. So often I am thinking about what to say or what my daughter is saying, that I totally lose track of what is really happening." In this way he would get caught in the content of the message, with his left-mode processing dominating his right-mode processing, and he would become blocked from awareness of the primary emotional sensations that created meaning and connection. While under water, the two of them were freed up from that left-side dominance, which allowed for a powerful sense of connection to be created through the exchange of nonverbal signals.

Around the same time, the girl said to her father, "Dad, you are becoming funny. Really funny!" Her father was very pleased with her observation. He said that he began to feel a new awareness of his body—sensations in his gut, feelings in his chest. He became more aware of images in his mind that weren't connected to particular word-based thoughts and described feeling "at ease" and connected to his family. He felt more in the present and less preoccupied with his own thoughts. With the growing connection of his right and left modes of processing he may be ready to explore the elements of his unresolved feelings surrounding the loss of his father and become aware of his feelings about his mother's and his colleague's illnesses.

Accepting our need for connection to others often requires the painful acknowledgment of how fragile and vulnerable we all are. This father's understanding of his experiences before therapy led to his awareness that his primary emotions had been shut off because of his early family life. An initial inborn appraisal that relationships are important and "good" had to be adaptively disconnected from the rest of his mind in response to his experiences with his family. This adaptation, a kind of minimizing defensive mechanism, was crucial to reduce the flood of feelings of longing and disappointment that would have incapacitated him in that emotional desert. That shutting-off process of his childhood was so efficient that it had blocked the elaboration of his appraisal/arousal systems from creating more intense or categorical emotions. He was left with a sense of numbness and a lack of meaning. His parents' distanced emotional communication had left him disconnected from the very processes that created meaning in his own life. With his work in therapy and self-reflective writing, his relationships as well as his emerging internal experiences could now reinforce a new integrated way of living that created more meaning in his life.

### EMOTIONAL COMMUNICATION

To "feel felt" requires that we attune to each other's primary emotions. When the primary emotions of two minds are connected, a state of alignment is created in which the two individuals experience a sense of joining. The music of our mind, our primary emotions, becomes intimately influenced by the mind of the other person as we connect with their primary emotional states. If communication is limited to a focus only on the rarely occurring categorical emotions, we lose the opportunity to experience the magical moments of connection that are available to us each day.

Resonance occurs when we align our states, our primary emotions, through the sharing of nonverbal signals. Even when we are physically separated from the other person, we can continue to feel the reverberations of that resonant connection. This sensory experience of another person becomes a part of our "memory of the other" such that the other person becomes a part of us. When relationships include resonance, there can be a tremendously invigorating sense of joining. This joining is not just in the moment: we continue to feel the connection to the other through the resonance of the relationship. Such resonance is experienced as memories, thoughts, sensations, and images of the other and our relationship with him or her. This ongoing sense of connection can be thought of as the ways our minds have become linked. This linkage reveals an integration of two minds.

So much of what happens in relationships is about a process of resonance in which the emotional state of one person reverberates in that of the other. Attuned connections create resonance. Another factor central to creating resonance may be "mirror neurons." The mirror neuron system is the finding that a particular kind of neuron directly links perception to action. This system of mirror neurons may be the early basis for how one mind creates the mental state of another inside itself.

Mirror neurons are found in various parts of the brain and function to link motor action to perception. For example, a particular neuron will fire if a subject watches an intentional act of someone else, such as the lifting of a cup, and will also fire if the subject herself lifts a cup. These neurons don't merely fire in response to any action seen in another person. The behavior must have an intention behind it. Waving hands in a random way in front of the subject does not activate a mirror neuron. Carrying out an action with an intended outcome does. In this way, mirror neurons reveal that the brain is able to detect the intention of another person. Here is evidence not merely for a possible early mechanism of imitation and learning, but also for the creation of mindsight, the ability to create an image of the internal state of another's mind.

Mirror neurons may also link the perception of emotional expressions to the creation of those states inside the observer. In this way, when we perceive another's emotions, automatically, nonconsciously, that state is created inside us.

For example, we may begin to cry when we see someone else crying. We learn how someone is feeling by putting ourselves "in the other person's shoes"—we know how others are feeling by how our own body/mind responds. We check our own state to know the state of mind of another person. This is the basis for empathy.

At the heart of feeling joined is the experience of empathic emotional communication. The way one mind becomes interwoven with another is through the sharing of the surges of energy that are our primary emotions. When children feel positive sensations, such as in moments of joy and mastery, parents can share these emotional states and enthusiastically reflect and amplify them with their children. Likewise, when children feel negative or uncomfortable sensations, such as in moments of disappointment or hurt, parents can empathize with their feelings and can offer a soothing presence that comforts their children. These moments of joining enable a child to feel felt, to feel that she exists within the mind of the parent. When children experience an attuned connection from a responsive empathetic adult they feel good about themselves because their emotions have been given resonance and reflection. Perhaps a story can bring more clarity to this concept of attunement.

## ATTUNEMENT AT THE SYCAMORE TREE

Mary observed this interaction between a young girl and a student teacher on the school playground. Sara was a very tentative four-and-a-half-year-old who was socially and physically cautious and was very hesitant to try new experiences. The teachers had worked carefully to build her confidence by providing learning opportunities that she could master with support and encouragement.

It was the spring semester and Sara was just beginning to challenge herself. In the playground was a large sycamore tree that had fallen over many years before and had been left as a natural bridge spanning ten feet of the yard. The children loved to walk across the log. Doing so was a great accomplishment. Sara, however, had never ventured to attempt walking on it until one day in mid-May, her confidence budding like the lilacs, she stepped onto the log and completed her journey to the other end. A student teacher had been watching her and as soon as Sara stepped off the end of the log the young woman exploded with cheers and applause because she was so excited about Sara's accomplishment. "Yea! Hooray! You're terrific! You're the greatest!" the teacher exclaimed loudly, jumping around and waving her arms with excitement. Sara looked at the teacher shyly and, standing rigidly, managed a faint smile. For weeks afterward, Sara avoided the log and it took a great deal of encouragement for her to try it again.

What was amiss in this interaction? Certainly the student teacher was positive toward Sara's accomplishment, but she missed attuning to Sara's experience. Her response reflected her own pride and excitement and was not about Sara's experience of mustering her courage and taking a great risk. The essence of Sara's experience was not reflected by the teacher's remarks. Actually they were overwhelming to Sara and did not support her going back to risk trying to walk on the log again. "Maybe I won't be able to do it that well again. I'd better not try it or I might fall off," Sara most likely thought. Being "terrific" is a hard act to follow for a cautious child, so she played it safe and did not build on her initial accomplishment.

What could the teacher have said that would have nurtured the emerging self-confidence that Sara was exhibiting? How could she have encouraged her in a way that would have been respectful of Sara's experience? How could the teacher's response have encouraged Sara to reflect on her own accomplishment in a way that she could own it and develop it further with confidence?

If the teacher could have reflected what she saw Sara accomplish in a warm and caring way, Sara could have seen herself reflected in the teacher's response. The teacher might have said, "Sara, I watched you carefully put one foot in front of the other and you walked all the way across to the other side. You did it! It was a little scary, since it was your first time, but you kept going. Good for you! You are really learning to trust your body."

These statements would have been the teacher's reflection of Sara's experience. The teacher's actual reaction tells more about the teacher's experience than it does about Sara's. The proposed reflective message she could have offered would have allowed Sara to integrate the experience and build on her accomplishment. This is an attuned, connecting response. This collaboration offers a reflection of both her outer behaviors and an understanding of her internal mental processes to provide an authentic reflection of Sara's ongoing experiences. In order to enrich coherent self-knowledge, children need to experience the integrating reflections of others in a way that matches both their internal and external experience. When we join with our children in attuned communication we support them in developing an integrated and coherent story of their own lives.

Attuned communication supports the emergence of a more autonomous self and flexible self-regulation. Emotional communication enables a form of joining that is truly an integrating process that promotes vitality and well-being in both parent and child. This experience of joining helps children develop a stronger sense of themselves and enriches their capacity for self-understanding and compassion.

This doesn't mean that we always have this joining experience, or that we have to be listening to and reflecting our child's experiences all the time. Continual intensity of a parent's focus on his child could actually be experienced as quite intrusive by the child. Within parent-child relationships there are cycling needs for connection and separation. It is important for parents to be sensitive to times when the child needs solitude as well as joining. The attuned parent respects the natural oscillating rhythms of the child's need for connection, then solitude, and then connection again. We are not designed to be in alignment all of the time. Attuned relationships give respect to the rhythm of these changing needs.

## WHY IS EMOTIONAL RELATING SO CHALLENGING?

Emotional relating requires a mindful awareness of our own internal state as well as being open to understanding and respecting our child's state of mind. We need to see a situation from our child's point of view as well as our own. When we are unaware of our own emotions or paralyzed by leftover issues and the emotional reactions that come from them, this can be very hard to do. As we've discussed earlier in the book, we often have leftover and unresolved issues from which we are reacting in "knee-jerk" ways that may immediately impair the opportunity for joining with our child in an attuned way.

Often we don't check in with ourselves and are unaware of our own primary emotional state. Even the categorical emotions may be denied and only brought to our awareness by behavior that projects our feelings outward, often in ways that are hurtful to our children. That is why it is important that we try to be aware of our own emotional processes and respect their central role in both our internal and interpersonal lives. Children are particularly vulnerable to becoming the targets of the projection of our nonconscious emotions and unresolved issues. Our defensive adaptations from earlier in life can restrict our ability to be receptive and empathic to our children's internal experience. Without our own reflective self-understanding process engaged, such defensive parental patterns of response can produce distortions in a child's experience of relating and reality.

For example, a recently divorced mother whose husband had left abruptly would become furious with her three-year-old son's demands for her time. Still unable to deal with her own sense of loneliness and abandonment, this mother found her son's "demanding behaviors," which expressed his need for connection to her, threatening. Her young son became the object of her anger as she projected her discomfort with her own neediness onto him. She felt rejected and alone, feelings not dissimilar to experiences in her own childhood, and she was unable to be receptive to her own child's needs for connection. This is how a child can become trapped in the emotional turmoil of a parent's unresolved issues. As this mother processed the pain of her divorce and integrated this significant life change into her understanding of her own childhood, her emerging life story enabled her to move through her defensive reactions and she no longer became unreasonably angry with her son. She came to realize how her own unresolved issues had been impairing her ability to respond warmly to her son's healthy desire to be close to her.

Self-reflection and an understanding of our internal processes allows us to choose a greater range of responses to our children's behavior. Awareness creates the possibility of choice. When we are able to choose our responses we're not being controlled by our emotional reactions that are often not directly connected to our children: they are overreactions driven more by our own emotional state than by engaged emotional communication with our child at the moment. The integration of our own self-knowledge facilitates our being open to the process of becoming emotionally connected with our children. Coherent self-knowledge and interpersonal joining go hand in hand.

When our internal experience keeps us from connecting with our children, their experience of our intense emotion may trigger the arousal of a defensive emotional state in them. When this takes place, we are no longer in a collaborative relationship but each person has separated into his or her own internal world and feels alone and isolated. When both the parent's and the child's authentic, inner self is hidden behind the mental walls of psychological defense, neither person feels connected or understood. When our children feel this sense of aloneness they may express their fear or discomfort at this disconnection either by behaving aggressively or by withdrawing. Children's behavior may then become the focus of our attention and our own feeling of isolation can keep us from making meaningful attempts at reconnecting with our children. In this way, our own emotional issues can create responses in our children that further impair our ability to emotionally understand them or ourselves.

Without emotional understanding it is difficult to feel connected. Emotional relating opens the door for collaborative, integrative communication in which a dialogue can take place that allows us to connect to each other. All relationships, especially parent-child relationships, are built on attuned, collaborative communication that gives respect to the dignity and uniqueness of the music of each individual's mind.

## INTEGRATIVE COMMUNICATION

Connecting to our children can be one of the most challenging and one of the most rewarding of experiences. As a parent we have the possibility of building lifelong meaningful relationships with our children when we learn to develop a sense of joining through integrative communication. When we align ourselves with them, we begin a process in which the basic elements of our minds become integrated. This

### TABLE 4. PRACTICES OF INTEGRATIVE COMMUNICATION

1. Awareness. Be mindful of your own feelings and bodily responses and others' nonverbal signals.

2. Attunement. Allow your own state of mind to align with that of another.

3. Empathy. Open your mind to sense another's experience and point of view.

4. Expression. Communicate your internal responses with respect; make the internal external.

5. Joining. Share openly in the give-and-take of communication, both verbally and nonverbally.

6. Clarification. Help make sense of the experience of another.

7. Individuality. Respect the dignity and uniqueness of each individual's mind.

linkage of minds enables us to have a vital sense of being with them. Integrative communication happens across the life of our relationship with our children. A sense of resonance begins to emerge through which we carry the other person with us even when we are physically apart. When our children experience this resonance with us it enables them to feel comforted even in our absence. They feel felt by us and can sense that they are in our minds, just as we are woven into their developing sense of self. This feeling of being connected gives children a sense of security and supports their exploration of their own emotions and the world around them. Emotional communication builds the foundation of our relationship with our children as well as their relationship with others.

We have found the practices shown on page 69 to be helpful in achieving communication that fosters a sense of joining.

Integration is a process whereby separate parts are linked together into a functional whole. For example, an integrated family enables members to be distinctly different from each other and encourages them to respect these differences while joining together to make a cohesive family experience. Communication in such an integrated family system has a vitality to it that reflects the high degree of complexity achieved by blending these two fundamental processes: differentiation (people being separate and unique individuals) and integration (people coming together to join with each other). Such a blending of differentiation and integration enables the family to create something larger than the sum of the individual parts. Integrative communication enhances both parents' and children's individuality as they join together as a "we" that enhances their sense of connection in the world.

## INSIDE-OUT EXERCISES

1.   Think of a time when you and your child had a different reaction to the same experience. Now try to see the
events from your child's point of view. How did you ap-
praise the meaning of the experience differently? How do
you think your child would react if you offered her a view
into how you have made sense of the experience through
her eyes?

2.   Consider the practices of integrative communication
as shown on page 69. Observe your interactions with your
child and think of how these seven elements may have
been a part of your communication. Try developing these
practices in your future interactions with each other. Can
your child sense being felt by you? How does your own
sense of joining with your child evolve?

3.   Think of ways that you can utilize these seven practices in communicating with your self. How can you experience an openness to your own internal states, your
primary emotions? Become aware of the inner sensations,
thoughts, and images that enable a deeper and centering
sense of mindfulness. As you let these internal processes

float into consciousness, empathize with them without judging or trying to fix yourself.

 SPOTLIGHT ON SCIENCE

### The Evolution of Emotion

The science of emotion is a complex field comprising the study of culture, mental processes, and brain functioning. Studies of animals suggest that humans are not the only beings that have feelings: animals' responses to danger evoke behavioral and physiological reactions of fight, freeze, or flee that appear to be quite "emotional." Within the animal kingdom, however, mammals seem to be a unique group in their evolution of a complex means of communicating their internal state to that of other members of the species. Emotion in this sense is a reflection of internal processes and can be communicated externally to others. Going along with this behavioral achievement is the evolution of the corresponding brain systems that mediate emotion and motivation: the *limbic* structures of the brain.

Paul MacLean has coined the term "triune brain" for his view of the brain as consisting of three separate but interconnected parts: a deep portion, the brainstem; a middle, limbic, section; and a higher region, the neocortex. The brainstem areas are considered "primitive" structures and are sometimes called the reptilian brain. As animals evolved, the next layer, the limbic circuits, became a part of the mammalian heritage. Limbic structures include the amygdala (mediates the important emotions of fear and probably anger and sadness); the anterior cingulate (serves as a kind of chief operating officer and performs executive control functions for the allocation of attention); the hippocampus (mediates explicit memory and serves as a cognitive mapper to give context to memory); the hypothalamus (the site of neuroendocrine processing linking body to brain via hormonal balance); and the prefrontal orbitofrontal cortex (integrates a wide array of processes, including emotion regulation and autobiographical memory).

Though the limbic region doesn't seem to have exact borders, its circuits do seem to share types of neurotransmitters and a common evolutionary heritage. The impact of limbic activity is widespread, going throughout itself and both down to the brainstem regions as well as up to the third area of the brain, the neocortex.

The limbic area not only regulates the brainstem but also can be said to coordinate internal states, including bodily functions, with interactions with the environment, especially the social environment. Indeed, the limbic region's activity helps to explain why most mammals are interested in the social world: social interactions help to regulate their own bodily functions! Coordinating the internal world of oneself with the internal world of others is a specialty of mammals and makes us exquisitely and inquisitively social beings.

As primates evolved within the large class of mammalians, the brain changed further. Evolution is not merely the addition of new structures; rather, it involves the adaptation of old circuits for new functions. Thus, the top layers of the limbic region, the orbitofrontal and anterior cingulate regions, are also considered part of the evolving neocortical part of the brain.

The neocortex is the most evolved in our species known as *Homo sapiens sapiens*. We humans are capable of highly abstract perception, conception, and reasoning because of the complex, intricate foldings of our neocortex. Neocortical processes, especially in the frontal regions, enable us to think flexibly, to reflect on abstract ideas such as freedom and the future, and to use words to describe these complex thoughts to others. Words themselves, though limited in their ability to symbolize exactly what we mean, enable us to distance ourselves from the constraints of physical reality. This freedom liberates us to create—or to destroy. Language is a powerful symbolic system that enables us to manipulate the world and others around us. Words also give us the ability to communicate across the boundaries of time and space that usually separate one mind from another. The Greek poet Aristophanes wrote, "By words the mind is winged." The neocortex enables civilization and facilitates the evolution of human culture.

One might think from this "triune" layering of the deep, limbic, and neocortical brain levels that the neocortex, the site of abstract reasoning and enabler of civilization, is the boss. But this is not so. At least, the situation is not so simple. Reasoning has been shown to be profoundly influenced by emotional and somatic—bodily—processes; the neocortex's activity is directly shaped by the neuronal processes of the limbic region and the brainstem. Emotions and bodily states influence reasoning. Our "higher," more evolved neocortex does not call the shots by itself. The social, emotional, and bodily processing of the other areas of the brain directly shape the abstract perceptions and reasoning of the neocortex.

## The Brain and Social Life

Almost all mammals have social ties, and those between mother and baby are especially strong. The limbic regions of the mammalian brain enable it to achieve a convergence of two fundamental processes: (a) maintaining the balance of more primitive body-brain functions such as heart rate, respiration, and sleep-wake cycles and (b) taking in information from the outside world—especially the social world of other mammals.

The amygdala is crucial in the perception and outward expression of facial responses and it is central in the regulation of emotional states. This dual inward-outward function also enables a unique mammalian preoccupation: we are focused on the internal states of others—especially parents on their offspring. This focus by parents on their offspring enables them to develop a balanced way of regulating their young's internal states. This is as true for rats as it is for monkeys and other primates, such as us. Lower animals, such as reptiles, amphibians, and fish, lacking in this evolved limbic circuitry, do not make the same fuss over their relatives' internal states, nor do they have the same form of emotionally attuned social life as mammals. In contrast, we can have emotional connections with other people, as well as with other mammals such as the dog that may be sitting by your feet right now.

## Neurons That Mirror Other Minds

All mammals have evolved limbic circuitry to "read" the internal states of others; in addition, it seems that primates have developed a unique capacity to create an internal state that resembles that of others. Primates respond to the mental states of others as they are revealed in their intentional external behaviors. Humans, with their unique development of the neocortex and language function, are not only attuned and aware of mental states, they are capable of assuming the perspective of the other person. We are beginning to unravel some of the neural mechanisms that may enable such incredible social functions to take place.

"Mirror neurons" were first discovered in monkeys in the 1990s and have been found more recently in humans. This fascinating recent discovery by neuroscientists has generated an exciting new set of questions into the nature of empathy, culture, and human relationships. In the Center for Culture, Brain, and Development at UCLA, Marco Iacoboni, one of the investigators who discovered mirror neurons in humans, and his colleagues are exploring some ideas about how mirror neurons may function in transmitting aspects of emotional and social life across cultures.

Mirror neurons are further hints at how our brains have evolved to be profoundly relational. As inherently social animals that have survived through evolution because of our ability to read each other's external expressions as signs of internal states, mirror neurons may enable us to readily and accurately respond to others' intentions. The survival value of being able to "read minds" in a social setting to determine another's status as friend or foe is profound. Hence, here we are with a legacy of empathy and the capacity for mindsight that is rooted in our evolution.

The ramifications of such a system in helping us understand social experience are powerful. Empathy can be seen as metaphorically putting ourselves in the other person's mental shoes. We may learn to understand others' internal states by the states our mirror neuron systems create inside us. Emotional understanding of others is directly linked to our awareness and understanding of ourselves.

For relationships between children and parents, it appears that adults who have mindsight, who can express their awareness of the internal events within themselves and in others, appear to have the coherence of mind that is correlated with having a child who thrives. We have no idea at this early state of knowledge how the mirror neuron system might be influenced by optimal or suboptimal parent-child experiences to enable children to grow well or to be impaired in their development. We also don't yet know the role such a system may play in the day-to-day empathic interactions between parent and child, or in the ways adults come to make sense of their own lives and the telling of their life stories. These are areas of research and future discovery in this exciting emerging field.

## Emotion as an Integrative Process

One idea emerging from various academic disciplines that explore emotion is that this elusive process has an integrative function. Some authors state that emotion links physiological (bodily), cognitive (information processing), subjective (internal sensory), and social (interpersonal) processes. Others write about the relationship between the regulatory and regulating aspect of emotion: emotions regulate the mind and are regulated by the mind.

Clinicians may find themselves thinking circularly when they try to describe emotion: emotion is what happens when you feel something strongly; feelings are generated when you are emotional.

It may be that the reason these researchers and clinical observers of human emotion run into such conceptual loops when they consider human emotions is that they all are describing portions of a larger picture—the old blind-men-and-elephant story. The big picture is that emotion has something to do with a process called neural integration. Integration is the linking together of separate components of a larger system. Neural integration is how neurons connect the activity of one region of the brain and body to other regions. Within the brain, regions called "convergence zones" have wide-reaching neurons that extend to distinct areas to bring together the input from a variety of zones into a functional whole. Zones that we've described earlier include the prefrontal cortex and the hippocampus. Convergence zones integrate neural activity in a range of areas. Other neural activity is integrated by the corpus callosum, the bands of neural tissue that relay neural messages between the brain's two hemispheres. The cerebellum may also serve to link widely distributed areas to each other. The amygdala, described above, also has extensive input and output fibers that interconnect various elements of perception, motor action, bodily response, and social interaction.

Viewing emotion as an outcome of shifts in neural integration enables us to consider the wide impact of emotion on the functioning of a person. It also enables us to see how emotional dysfunction may occur when the normally integrating processes of the brain that enable balanced functioning have been impaired. Dis-integration is the result with a tendency toward states of rigidity or chaos. On the interpersonal level, we come to feel emotionally connected when our

minds become integrated with each other. These linkages of one mind to the other occur when the subjective, internal state of one person is respected and responded to by the other. We feel how our minds exist in that of the other person. This can also be seen as the integration of the activity of two brains: neural integration, interpersonal style.

## Digging Deeper

L. Ciompi. "Affects as Central Organizing and Integrating Factors: A New Psychosocial/Biological Model of the Psyche." *British Journal of Psychiatry* 159 (1991): 97–105.

A. Damasio. *Descartes' Error.* New York: Simon & Schuster, 1994.

———. *Self Comes to Mind: Constructing the Conscious Brain.* New York: Random House, 2010.

B. L. Fredrickson. *Love 2.0: How Our Supreme Emotion Affects Everything We Feel, Think, Do, and Become.* New York: Hudson Street Press/Penguin, 2013.

D. Fosha, D. J. Siegel, and M. Solomon, eds. *The Healing Power of Emotion.* New York: W. W. Norton, 2009.

J. J. Gross and R. A. Thompson. "Emotion Regulation: Conceptual Foundations." In J. J. Gross, ed., *Handbook of Emotion Regulation*, pp. 3–26. New York: Guilford Press, 2007.

M. Iacoboni. *Mirroring People.* New York: Picador, 2009.

———. "Imitation, Empathy, and Mirror Neurons." *Annual Review of Psychology* 60 (2009): 653–670.

J. LeDoux. "Rethinking the Emotional Brain." *Neuron* 73, no. 4 (2012): 653–676.

M. D. Lewis. "The Emergence of Human Emotions." In M. D. Lewis, J. M. Haviland-Jones, and L. F. Barrett, eds., *Handbook of Emotions,* third edition, pp. 304–319. New York: Guilford Press, 2008.

M. D. Lewis and R. Todd. "The Self-Regulating Brain: Cortical-Subcortical Feedback and the Development of Intelligent Action." *Cognitive Development* 22, no. 4 (2007): 406–430.

J. Panksepp and L. Biven. *The Archaeology of Mind: Neuroevolutionary Origins of Human Emotions.* New York: W. W. Norton, 2013.

E. T. Rolls and F. Grabenhorst. "The Orbitofrontal Cortex and Beyond: From Affect to Decision-Making." *Progress in Neurobiology* 86, no. 3 (2008): 216–244.

D. J. Siegel. *The Developing Mind, Second Edition: How Relationships and the Brain Interact to Shape Who We Are.* New York: Guilford Press, 2012. Chapter 4.

D. J. Siegel and T. P. Bryson. *The Whole-Brain Child.* New York: Random House, 2011.

R. Thompson, M. Lewis, and S. Calkins. "Reassessing Emotion Regulation." *Child Development Perspectives* 2, no. 3 (2008): 124–131.

# How We Communicate:
# Making Connections

## INTRODUCTION

L earning to communicate and listen empathically is a vital part of parenting. At the heart of this openness is *parental presence*— the receptive state in which we take in the signals from others and have compassion and kindness in our interactions. Caring communication supports the development of a healthy attachment that is especially important in building a trusting parent-child relationship. Studies performed across many cultures suggest that a common element in healthy attachments is the ability of the parent and child to have a reciprocal give-and-take of signals. This is called contingent communication and means that the signals sent by the child are directly perceived, understood, and responded to by the parent in a dance of communication that involves mutual collaboration. Parents feel good and their children feel good, too, when interactions are respectful and responsive to each individual. This contingent communication enables a vitalizing sense of connection that may be at the heart of nurturing relationships across the life span. This is how we "feel felt" by another person, the gateway to how a "me" joins with another and becomes a part of a "we."

Collaborative or contingent communication allows us to expand our own minds by taking in others' points of view and seeing our own point of view reflected in their responses. From the beginning of life, the infant requires collaborative communication in order to thrive. When a baby smiles and makes soft wordless sounds, a nurturing parent responds in like manner by smiling back at the baby and imitating some of the sounds, then pausing and waiting for the baby to respond again. A dialogue is begun which says to the baby "I see you and I'm listening to you and I'll give back to you a reflection of yourself that is valued so you can see and value yourself too. I like you just the way you are." Thus, a connection is made through this simple dialogue, a give-and-take of signals that creates a sense of joining. Our child's emotional well-being is built on this intimate dance of communication.

In contingent communication the receiver of the message listens with an open mind and with all his or her senses. Her reaction is dependent on what was *actually* communicated, not on a predetermined and rigid mental model of what was expected. The exchange occurs in the present moment without preoccupation with internal events of the past. Contingent communication is full of possibility for connecting because instead of responding in a rote manner, a parent responds to the signals actually sent by the child. In contingent communication the parent gives respect to the act of listening. So often parents don't listen to the message their child expresses because they are preoccupied with their own thoughts and feelings. A child's true message is not always immediately apparent and a parent may have to "decode" the message in order to understand it. It's important to remember that even if your children's messages don't immediately make sense to you, they are trying to get their needs met in the best way they can at that point in time.

As an example let's look at the scenario of a mother returning home after work and being met by her twenty-two-month-old son who runs enthusiastically to greet her as she enters the door. He wants to reconnect with her after their day of separation. The mother, however, has a different agenda; she wants to shift out of her professional role and into her mother role, so giving her child a quick, distracted hug she heads for the bedroom to change her clothes, saying, "I'll be back in just a minute." This brief connection followed by a separation is not satisfying to her child, who follows after her crying and demanding to be picked up. The mother tries to put the child off so that she can complete her own goal before attending to her child. The child becomes more upset, cries louder, lies down on the floor, and starts kicking his feet against the wall. This annoys the tired mother, who doesn't want to hear the banging or wash the scuff marks off the wall. The mother feels that her child is being unreasonable *and* demanding. She says sternly, "I'm not going to play with you unless you stop that kicking right now!" On hearing this, the child experiences an even greater disconnection because of his mother's anger toward him. He now becomes even more upset and takes a swing at his mother. Now the mother doesn't want to give her child any positive attention because in her mind he's not behaving well and the mother doesn't want to reinforce this "bad behavior." The child's message of how important it was to reconnect with his mother after a long day of separation wasn't received and he began acting out of frustration at not being understood. He then continued to seek and get connection, albeit through negative ways.

If the mother had understood the initial signal, she could have chosen to sit down on the couch for a few minutes to talk and cuddle with her child or read a book to him before going in to change her clothes. Knowing that reconnecting after a separation is very important to a young child, the parent can create more realistic expectations and avoid unnecessary frustration for both herself and her child. Contingent communication could have led to a different choice on the part of the parent that might have changed the parent-child interaction and, with both parent and child feeling connected, could very likely have altered the events of the rest of the evening. When a child doesn't feel understood, little things can become big issues.

## CONTINGENCY AND COHERENCE

Why is contingent, collaborative communication so important?

From a biological perspective, the way interpersonal communication shapes the neural structures from which a sense of self is created can be described as follows: When we send out a signal, our brains are receptive to the responses of others to that signal. The responses we receive become embedded in the neural maps of our core sense of self. A neural representation of the self-as-changed-by-the-other is created within our brains that becomes a central aspect of our sense of identity. If others' responses to us are contingent, our neural machinery creates an internal sense of the coherence in the connection between our self and the other person. There is a coherent relationship between the self before the signal was sent and the self after the signal was responded to.

How does this occur?

A contingent response is when the quality, intensity, and timing of the other's signals clearly reflect the signals that we have sent. With contingent interpersonal interactions, we create a neural sense of grounding and empowerment in a social world of connections. These kinds of connections create a strong, internal coherence of the self.

When contingent communication is present in our interactions, our sense of self with that person feels right. It feels good. We feel understood. We have a sense that we are not alone in the world, because our self is connected to something larger than the boundaries of our own skin. Over time, repeated patterns of contingent communication also enable us to develop a coherent autobiographical self that connects the past, present, and anticipated future. Both the here and now and the reflective autobiographical forms of conscious awareness shape our experience of ourselves in the world.

Collaborative, contingent communication allows us to expand our sense of self. As we feel connected to our children, we can become more receptive to them. This is the essence of collaborative communication. The responses of others are not merely mirrors of our own signals but incorporate the essence of the other person's view, which makes sense of our communication. In this manner, children come to feel felt: they come to feel as if their mind exists in the mind of their parent.

Learning happens in a social context. Through this joining together, children construct both social knowledge and an understanding of the self that could never have been created in isolation. This process is called *co-construction.* Self-knowledge is in part a co-constructive process as children learn about themselves through the communication and connections they have with their parents and others.

Let's look at the kind of experience that an infant who is crying because she has a wet diaper might have with her parent. In an ideal situation, the parent hears his baby crying and within a reasonable amount of time makes sense of this signal of distress, and responds by changing the child's diaper. The child experiences (1) discomfort with her wet diaper and calls out with a cry; (2) having a parent soothe the source of her distress; and (3) a sense of the self as changed by the interaction with the parent that is coherent. This is how interpersonal contingency creates internal coherence.

A different situation occurs if the parent does not perceive the baby's signal, or cannot make sense of it. The parent may not offer an appropriate response. For example, the parent may try to entertain the baby by playing with her, offering her food, or trying to rock her to sleep. The baby may experience (1) discomfort and crying; (2) noncontingent and noncomforting responses as she remains in a state of distress and disconnection from the parent; and (3) no coherent sense of "self as changed by the interaction with the parent." The baby is left in a state of isolation. The outer world does not provide comfort and the baby's sense of self becomes constructed from this disconnected form of communication. In experiences that are inconsistent, the child doesn't know what to expect or to depend on. The child then accumulates a "sense of self as changed by the interaction with the parent" as unreliable and inconsistent. Sometimes the contingent communication creates a coherent sense of self; sometimes it leaves the child in a state of isolation lacking in coherence. The resultant feeling in the child is that the world is not a reliable place and the sense of self may become filled with anxiety and uncertainty.

In order to thrive, a child (and perhaps each of us at any age) needs contingent communication with significant others. A child especially needs a "good-enough" parent: no parent can provide contingent communication all of the time, but frequent experiences of feeling connected are vital in building relationships. It is often a challenge to make sense of our children's signals and some children may be especially difficult to understand and to soothe. When the inevitable disconnections and misunderstandings occur, we can repair those ruptures so our children learn that a healing reconnection is possible. The accumulation of our children's positive, connecting experiences with us, through contingent communication and essential moments of interactive repair, enable them to build a coherent sense of self as they develop.

Being open to the complex process of contingent communication requires a willingness to become a part of something much greater than a self that is defined by the body's physical boundaries. Such intimate collaboration may be difficult when parents did not experience a sense of connection in their own childhoods. It is a challenge to remain open to the basic processes of perceiving, making sense of, and responding to the communication sent by our children. A child's experience of inappropriate, noncontingent communication can create emotional distress and limit her openness to making further attempts to connect. It is confusing to children if their reality of an experience is denied or misunderstood by their parent or another significant adult, because those are the very people with whom they most need to connect.

## DENYING REALITY

Every day we miss opportunities for making true connection because instead of listening and responding appropriately to our children we respond only from our own point of view and fail to make a connection to their experience. When our children tell us what they think or how they feel, it is important to respect their experience, whether or not it's the same as our own. Parents can listen to and understand their children's experience rather than tell them that what they think and feel isn't valid.

The following examples may help to illustrate these ideas. Imagine that your child is riding his tricycle and falls off. It looks to you more like a surprise than an injury, but he starts crying, to which you respond, "You're not hurt. You shouldn't cry. You're a big boy." Your child feels hurt, whether it is his body or his pride, and yet you tell him that his experience isn't a valid one. Now consider how the child might feel if you gave a contingent response: "It surprised you when you went over that bump and you fell off onto the grass. Are you hurt?"

Or let's imagine that your child enthusiastically expresses a desire for a particular toy that she has seen advertised and you respond with,

"Oh, no, you don't really want that—it's just a piece of junk." Your child just told you that she does want it, which doesn't mean that you have to get it for her, but you can acknowledge her desire. "That toy really looks like it would be fun to play with. Tell me what you like about it." If she continues to insist on getting the toy right away, you can say, "I see that it is hard to wait when you like it so much. Maybe you want me to write it down so when it is time to get a present, I'll know what you might like to have." When parents understand that they can let their children have and express their desires without having to fulfill them, it frees the parent to make a connection to the child's experience without having to deny their feelings.

Here's another more extreme example of noncontingent communication that Mary observed while visiting a kindergarten class. The teacher was working with a small group of children while the rest of the pupils were working at their desks independently. A young boy was having some difficulty with his independent project. After some time struggling alone, he took his paper and went to seek help from the teacher. Not wanting to interrupt her, he stood quietly for a few moments, hoping to be noticed. The teacher ignored him, since she had instructed the children to work at their desks and he was not at his desk. He then spoke to her to get her attention in order to ask his question.

Without turning her head to look at him, she said, "Andy, you are not here."

Andy looked puzzled. After waiting a moment, he touched her shoulder and then repeated his question.

She then turned her head, looked straight at him, and said: "Andy . . . you are *not* here!" She was focusing on her rules rather than on his communication.

Andy turned away. With downcast eyes and head hung low, he

slowly made his way back to his seat. He sat down and halfheartedly made some marks on his paper.

Andy had been frustrated with his work and was seeking connection and help. What was his internal experience of that interaction with the teacher? For any of us, not having connection when we need it can induce intense emotion. That emotion is shame. For a five-year-old, trying to understand the words "You are not here" must have also been quite confusing. He not only has noncontingent communication to deal with, but the teacher's response is "crazy-making"—her words deny both his reality and also her own behavior! If he were truly "not there," why was she even talking to him? This is a prime example of a mismatch among her words, her behavior, and his need for connection.

When there is a mismatch between a child's need for connection and a significant adult's indifferent responses, the child can feel isolated and alone. When a child's emotions are activated, the child often needs connection. At those moments of increased need, children are most vulnerable to insensitivity.

## COMMUNICATION ACROSS THE HEMISPHERES

Communication includes the components of both verbal and nonverbal language. The nonverbal component of communication helps us to feel connected and grounded. Being understood takes more than words. Nonverbal messages are often perceived unconsciously, and these signals have a deep and moving effect on how we feel. One way of understanding this is that the right side of the brain is designed to both send and receive nonverbal signals, and is also dominant for regulating our internal emotional states. How nonverbal signals are contingently shared may have a profound impact on how our minds create a state of balance. The left hemisphere of the brain specializes in sending and receiving verbal data. This means that we may have word-based thoughts that are quite distinct from our nonverbal internal sensations. The signals sent from a person's right hemisphere directly shape the activity of another's right hemisphere. The same is true for the left: words from others' left hemispheres activate our own left hemispheres. If the verbal and nonverbal signals sent from another person are congruent, then the communication can make sense.

If verbal and nonverbal signals communicate different messages—are not congruent—the overall message will be unclear and confusing. We are getting two different and conflicting messages at once. Suppose a mother is sad and her daughter, picking up the nonverbal signals, asks, "Mommy, what's wrong? Did I do something to make you sad?" and with a forced smile, her mother replies, "Oh no, honey, I'm not sad, everything is just fine." The child will feel confused because of the double message. Her experience is informing her of one thing while the words of her mother are giving a contradictory message. If there is a mismatch between the verbal and the nonverbal, it can be quite disorienting for a child trying to sort out the confusion and the incoherence of the communication.

We may have learned in our own childhood that emotions are "bad" and therefore our children's as well as our own emotions may be uncomfortable for us. Such experiences and such beliefs may block our ability to be present fully in our relationships. Our children benefit when we express our feelings directly, simply, and in nonthreatening ways. A child wants to know not only what his parents think but also how they feel. When we're upset, angry, or disappointed, excited, proud, or delighted, we can let our children know. Children need to know that we have feelings too. When we express our emotions, our children learn what is important to us as well as witness a model for the healthy expression of emotion. Children learn to be empathic

## TABLE 5. COLLABORATIVE COMMUNICATION

PROCESSES OF COMMUNICATION
Receive—Process—Respond

PATHWAYS TO COLLABORATION
Explore—Understand—Join

PATHWAYS TO DISCONNECTION
Interrogate—Judge—Fix

through seeing the ways in which we emotionally respond, not just in the words we say. We can be authentic and passionate as we respect both our children's and our own experiences.

Our sense of self becomes defined by the contingent ways in which we connect with others. Our brains are structured to be connected to other brains. Collaborative communication involves the spontaneous connection of each side of the brain to that of the other person as we share signals in both the verbal (left) and nonverbal (right) domains. This dance of communication not only enables us to feel close and connected to others but also allows our minds to feel coherent and in balance. Our sense of "I" is profoundly influenced by how we belong to a "we."

### OPENING THE CHANNELS OF COMMUNICATION

How do we become present, more open to collaborative communication? To clearly communicate with our children and others we need to receive, process, and respond to the message that was sent.

Receiving a message in both its verbal and nonverbal forms is the first part of the communicative process. Verbal signals include words that describe our ideas, thoughts, feelings, and any entity that can be translated into words. These come from the left side of our brains. Non-verbal signals include eye contact, facial expression, tone of voice, gestures, posture, and the timing and intensity of response. These are both sent and received by the right side of our brains. Often, the emotional, meaning-making aspects of communication come primarily from our right hemispheres. It is crucial to pay close attention to the nonverbal component of all communication. Through the sharing of these nonverbal signals deep connections are created between two individuals.

Processing enables us to make sense of the signals we perceive and choose a contingent response. Processing involves filtering complex layers of assessment through the lens of our mental models, which have been shaped by our past experiences. These internal processes influence how we interpret the signals received in the present as well as how we anticipate the future. Our responses will be shaped by how we have received and processed the signals and the meaning we bring to the message and the way it was communicated.

When we listen to our child's signals we can learn more about his state of mind and point of view. Our understanding of our child is of primary importance as we internally process the signals we have received. This internal work of processing also involves our own assessments of the experience. A true collaboration involves the blending of both minds, which gives respect to an understanding of both our own and our child's experience. If parents only understand their own experience and don't connect with their child's experience, they will very likely have difficulty in developing a close and meaningful relationship with their child. On the other hand, if parents only consider their child's point of view and neglect their own internal experience they will very likely have difficulty in setting boundaries with their child. Parents may begin to resent their child for being too demanding. If parents forget to consider their own internal state and take into consideration only the needs and desires of their child, it can leave the parent feeling angry and exhausted and the child feeling insecure from a lack of boundaries.

Healthy relationships require that we make choices that support our children's need for love and nurturing and create experiences that bring structure to the complex dynamic of the parent-child relationship. For example, a mother was getting ready to cook dinner for company when her five-year-old daughter came into the kitchen wanting to work with a watercolor paint set at the kitchen counter. Now, even though the mother valued her daughter's creative experience and her ability to self-initiate activities, the timing of this activity would greatly interfere with the mother's ability to continue preparing for her guests. If the mother let her child paint at the counter, she would probably end up feeling irritated with her. An abrupt "no" response may have created frustration for her daughter, and a disconnection and possible argument that would needlessly require time and energy. A contingent response could be something like, "I know that you love to paint but I am really busy getting a dinner ready for our company and I'm afraid that I'd get cranky having you painting here in the kitchen." A contingent response would allow the mother to join with her child in a collaborative process of communication to create an outcome that would be of mutual satisfaction to both mother and daughter.

A contingent response is not just a mirror reflecting back to the other person an exact replica of the sent signals. Such a mirroring can be extremely frustrating. "I'm so mad that I can't go to the park!" a boy said to his mother. Her mirroring response "You feel mad that you can't go to the park" led him to cover his ears and stomp out of the room. Instead, a collaborative response might go something like this: "I know you really want to go to the park today. I wish we could go too. It can be disappointing and frustrating when plans have to change." This response reveals that the mother has received his signals, has processed them in a way that shows that she understands his state of mind, and shares them in a way that reflects not only what he has said, but also the fullness of his emotional experience.

In noncontingent communication we may often interrogate, judge, and then try to fix a situation. Interrogation involves the aggressive questioning of another with presumptions about what the other person may be experiencing and with ulterior motives in seeking out particular responses. For example, if your shy ten-year-old daughter has been having difficulty making friends at her new school, you may become anxious about her social relationships. When she comes home from school, you might immediately greet her with questions relating to her friendships, such as, "Did you play with anyone today?" or "Did you talk to the girls at lunch?" While your intent may be to help your daughter, asking such intense questions of someone who is already feeling insecure about her social life may make her feel even more inadequate.

Judging makes assumptions about "right or wrong" aspects of the other person's experience. We may be critical of another person's different approach even as we are trying to receive her signals. Sometimes these judgments come from rigid mental models that we have within ourselves. Often, we may be unaware of these internal models and the biases they create that keep us from being open and receptive. For example, because you wish your daughter could be more outgoing, you may reveal your disappointment in her through your questions and your behavior. You may give the message, indirectly through nonverbal signals of disapproval or directly through the words you choose, that she is defective in some way. "If only you would be nice to your friends, I'm sure they'd want to play with you more." Or "Why don't you act like your cousin Susie? She's always friendly." These judgmental statements do not help your child feel understood or supported and do not boost her self-confidence.

If our responses are intended to quickly fix a situation, we lose the opportunity to join with our children in collaborative communication. Also, trying to fix our children's problems does not give respect to their own ability to think and figure out solutions to their own difficulties. Of course, parents are an important support in helping children to learn problem-solving skills. But automatically trying to fix things before we have joined with our children's experience can be both intrusive and disrespectful. In the example above, inviting several girls over without your daughter's involvement in the process could be intrusive and might not work out well. Instead, being open to the process of how your daughter experiences her relationships with others might reveal new insights into how you can understand her difficulties and support her learning new social skills that would bring her more satisfaction in her peer relationships. Knowing and accepting her temperament will enable you to offer her support while at the same time nurturing her ability to gradually develop her courage and reach out to others. With your attunement and understanding, she can become more secure. Feeling supported by you, she can brave the world with more strength and willingness to try new things. Instead of trying to fix, think of trying to join. Be sure to keep an open mind as you try to understand your children's perspectives.

## PROCESS AND CONTENT

Becoming aware of both the process and the content of our interpersonal communication is a fundamental part of coherent self-knowledge. So often we focus on the content of what is being said and lose track of the process of connection. The meaning of our interactions, however, is often found in the process, not merely the content. What does this mean? Communication is how we engage with others in the process of connection, not only the sharing of particular informational content. The dynamic flow of information, the ways we send signals back and forth, connects us to each other. As we engage in the process of communication—the exchange of the energy and information that are the essence of our minds—we connect with one another.

When parents have leftover or unresolved issues, they often project this baggage onto interactions with their child. The child's signals are then filtered through a rigid lens of the parents' model of the world, the distortions of their closed and unreceptive minds. When parents are locked into their own point of view as the only point of view, they shut down their channels for open collaborative communication.

If collaboration within the parent-child relationship is broken, the child's mind may close down the channels of communication and no longer be receptive to learning. When we're not in connection with our child it is extremely unlikely that any supportive communication will take place and there is a high likelihood that both parent and child will end up frustrated, angry, and feeling more distant and isolated from each other.

Our lives reflect both the process of connection and the content of what happened to us in the past. Our families of origin have shaped not only what we remember about our childhoods, but also how we have come to remember our lives and create a coherent mind. The give-and-take of contingent collaborative communication may not come naturally because it was not a part of our own childhood. Fortunately, though, we can *learn* to listen to our child and become aware of his perspective as well as our own. Parents' patterns of communication shape the coherence of their child's mind. Becoming aware of both the process and the content of our interpersonal communication is a fundamental part of coherent self-knowledge.

## INSIDE-OUT EXERCISES

1. Think of an experience from your own childhood where your reality was denied. How did it make you feel? What was happening to your relationships with your parents during that experience?

2. Observe others' conversations. First, focus on the words being said and the factual statements being expressed. Next, notice the tone of voice used by the communicators. Are the verbal and nonverbal aspects of spoken language congruent? How do the other aspects of nonverbal communication mesh with the language used? How would you feel in that situation?

3. Notice when two people don't keep the channels of communication open. How do you perceive the distance between them? Is an argument imminent? Think of your own patterns of communication. In what ways do you interrogate, judge, or fix? The next time you are communicating with your child, try to explore, understand, and join. What do you notice with this approach? How do you think your own childhood experiences with these forms of communication have affected your ways of relating to others?

### The Importance of Contingency

Old views of the nature of the human brain envisioned a neural system that could live within a body and function as an independent structure. This view examined the brain in isolation, as a fascinating structure worthy of active study into its mysteries. With the emergence of technological advances, we can now peer into the functioning of the living brain. As the mysteries of this extremely complex organ have begun to unfold, we are now realizing how profoundly relational our brains are. Our brains are constructed to be directly influenced by their interactions with other brains. This is not a fluke of nature but it is an outcome of evolutionary factors that have favored a plastic—changed by experience—and highly social organ that enables us to be influenced by and influence our companions.

The relational mind can be understood from the basics we've discussed: the mind emerges as the flow of energy and information occurs within a brain, or between brains. Studies of infants have long revealed the interpersonal nature of our earliest days: attuned, reciprocal, mutual, collaborative, contingent communication describes the fundamental universal process that links infant to parent. Researchers wanting to know more about this process devised ingenious approaches to studying parent-child connections.

Colwyn Trevarthen, a research psychologist originally trained in brain science and now one of the world's leading early-infancy researchers, knew that appropriate responses of a parent were described as being essential for an infant's well-being. In Ed Tronick's "still-face experiment," a parent is asked to maintain an unexpressive face as a young child attempts to communicate. Initially the infant responds with increased attempts to connect, then protests with agitation, becomes disorganized, and then withdraws. Researchers had interpreted this as the child's need for attuned communication. But Trevarthen realized that this experiment didn't rule out the possibility that children merely needed a positive response, not necessarily an attuned, contingent one. To address this issue, and to clarify once and for all what was the essential ingredient in parent-infant communication at this early age (the infants were three to four months old), he and his colleagues devised the "double television" experiment.

Infants love to peer into the faces of their parents. Imagine a closed-circuit television system where the infant peers into a monitor that has the face of his mother on it. An ingeniously angled mirror enables a camera to be focused right on the face of the infant. In other words, the infant is looking straight into the face of the mother on a monitor screen while also, unknowingly, looking into the lens of the camera. The camera then sends the image to another online monitor into which the mother is looking. She has the same setup: looking into her baby's face while simultaneously looking into a camera lens that sends her image back to the baby.

The first phase is to have them interact in real time, but looking at the monitor images. What one sees is identical to the face-to-face situation: contingent communication in which there is a sharing of nonverbal signals with the crescendos and decrescendos of energy reflecting alignment of their primary emotional states. This attunement reveals how the internal states of the interacting pair align themselves through nonverbally expressed affects. The second phase of the study reveals the importance of contingency. The videotaped recording of the mother's face from the prior minute was displayed to the baby. Now the infant was looking at his mother's face, seeing the same positive responses as before, but with one difference: they were no longer contingent because they occurred at an earlier time. Given that this was the tape of the last minute being replayed on the screen, the signals were still energized and positive—but now they did not correspond to the infant's own signals.

What do you think happened?

The infant had the same series of reactions as in the "still-face" situation, when the mother shows *no* facial responses: the baby got restless, agitated, disorganized, and then withdrew. This study clearly showed that infants need more than just the upbeat responses of their parents. They need those connections to be contingent.

We likely need the same contingency throughout our lives. Our sense of self is created within the relationships we have with each other. Contingent communication enables us to feel coherent, to neurally create a core self that is full and alive, and to go out in the world with a sense of vigor and vitality.

With ruptured communication, contingency is halted. At its most basic level, movement toward repair requires that contingency be reestablished. With reconnection, two internal states are realigned and the self of each person reaches a renewed sense of coherence.

### Contingency and a Coherent Self

The idea that communication with significant others shapes who we are is not new. A Russian psychologist named Lev Vygotsky wrote in the 1920s that thought is internalized dialogue. How we come to talk to ourselves is shaped by how others have talked with us. Those who study narrative as a central feature of how we come to define ourselves hold a similar view: we construct the narrative of our lives based on the nature of the interactions we have had with others. The child psychiatrist Daniel Stern, drawing on research on infant development, beautifully described how interactions between parent and child shape the development of the self during the early years of life.

Recent advances in brain science have suggested that our brains are exquisitely social. Evolutionary biology provides us with a view of the brain as the social organ of the body. We have hearts to pump our blood, kidneys to filter it, stomachs to digest our food, and brains to coordinate the internal world with the external world. Because we have evolved in groups, our survival has required that we be able to read the signals of others. This mind-reading process not only gives us information—it shapes us. Developmental studies of a process called "social referencing" point to the powerful ways in which the emotional responses of parents, as revealed nonverbally in their facial expressions and gestures, in an ambiguous situation determine a child's own emotional responses and behaviors.

Other recent studies in brain science have utilized the data from patients with brain lesions to explore ways of understanding how a sense of self is created, and how we become conscious. The neurologist Antonio Damasio has written about how specific areas of the brain of an individual create a sense of self. His work is done on individual patients and their neurological problems, but nevertheless we can build on this work to deepen our understanding of the relational mind and the social brain. The important point is that the brain neurologically creates a core self that incorporates the stimuli from the outside world. If we apply the idea of communication as the "stimulus" that changes the self, then we can envision how contingent communication can create a coherent core self. We propose that a coherent self is neurologically created when the response of the social environment is contingent.

This proposal fits well with Vygotsky's and Stern's notions of how our complex sense of self is built upon how experiences have impacted us. Consciousness of an autobiographical self is similar to what others have described as a past-present-future form of consciousness and is known as "mental time travel" in the writings of Endel Tulving and his colleagues. Our awareness of the moment-to-moment events that impact us is called primary consciousness and has been described by several researchers as a form of here-and-now awareness.

These convergent views suggest that our self is neurologically created in layers of circuitry that embed both our momentary interactions with a dynamic world and the accumulation of experience as embedded in our various forms of memory. As memory itself is reshaped by recollection and by the process of creating new and ever-evolving neural connections, so too does our sense of self remain open to growth and development. Such development of a coherent sense of self may involve the interactions of our here-and-now self experience with others as well as our mental time travel that enables us to create a past-present-future integrating and reflective autobiographical self as we live in dynamic and ever-interacting internal and social worlds.

* * *

### Integrating the Hemispheres

Connecting to each other involves at least two basic forms of communication: nonverbal and verbal. Our right hemispheres, the nonverbal, somatic, emotional, social, autobiographical processors, enable us to utilize nonverbal signals in communicating with one another. Science points to an interesting finding: the right hemisphere appears more directly linked to the limbic circuitry that mediates emotion and motivation. In this way, the distinction between right and left modes of processing likely includes the distinction between the emotional/limbic processing of the right hemisphere and the more intellectual, rational—sometimes called "neocortical"—processing of the left.

This distinction is particularly relevant to raising children for a number of reasons.

- Early in life, during the first year or two, the infant is primarily a right-hemisphere creature. Knowing how to use your own right hemisphere is crucial to making connections with your young child.

- As the child grows into preschool age, the corpus callosum, the bands of tissue connecting the two hemispheres, is quite immature. This condition leads children at this age to have an innate difficulty "putting words to their feelings." Sometimes their right hemispheres may be so intensely reacting that they enter a temper tantrum, a state in which they cannot use language to communicate. Nonverbal soothing communication may be best in these situations.

- Schools generally emphasize left-hemisphere processing over right-hemisphere. Adults, having been raised in those schools, often collude to reinforce this unfair bias toward words and logic. It is difficult for right-mode processing to make a case for itself without resorting to the logic of words for persuasion. Someone has got to stick up for the right hemisphere! Remember, the right-hemisphere brain processes are important for self-regulation, a sense of self, and empathic connections to others. Finding ways to encourage your child's right-hemisphere development and its integration with the left may be crucial for developing a sense of resilience and well-being.

## Digging Deeper

L. Cozolino. *The Neuroscience of Human Relationships.* New York: W. W. Norton, 2006.

D. J. Siegel. "Toward an Interpersonal Neurobiology of the Developing Mind: Attachment, 'Mindsight' and Neural Integration." *Infant Mental Health Journal* 22 (2001): 67–94.

————. *The Mindful Therapist: A Clinician's Guide to Mindsight and Neural Integration.* New York: W. W. Norton, 2010.

————. *The Developing Mind, Second Edition: How Relationships and the Brain Interact to Shape Who We Are.* New York: Guilford Press, 2012. Chapter 8.

D. Stern. *The Interpersonal World of the Infant.* New York: Basic Books, 1985.

C. Trevarthen. "The Self Born in Intersubjectivity: The Psychology of Infant Communicating." In U. Neisser, ed., *The Perceived Self: Ecological and Interpersonal Sources of Knowledge,* pp. 121–173. New York: Cambridge University Press, 1993.

————. "Lateral Asymmetries in Infancy: Implications for the Development of the Hemispheres." *Neuroscience and Biobehavioral Reviews* 20 (1996): 571–586.

————. "What Is It Like to Be a Person Who Knows Nothing? Defining the Active Intersubjective Mind of a Newborn Human Being." *Infant and Child Development* 20, no. 1 (2011): 119–135.

E. Tronick. *The Neurobehavioral and Social Emotional Development of Infants and Children.* New York: W. W. Norton, 2007.

# CHAPTER 5

# How We Attach: Relationships Between Children and Parents

## INTRODUCTION

Babies are born into the world dependent upon their parents for their very survival. It is most often the mother, but can be another loving, sensitive caregiver, who initially provides food and comfort for the newborn and to whom the infant develops a primary attachment. The baby and mother experience an intimate connection that gives the infant a sense of security. For infants, having a primary adult who is caring for them in sensitive ways, one who can perceive, make sense of, and respond to their needs, gives them a feeling of safety. The sense of well-being that emerges from predictable and repeated experiences of care creates what the attachment theory pioneer John Bowlby called a "secure base." This internal model of security enables children to develop well and explore the world around them. Secure attachment is associated with a positive developmental outcome for children in many areas, including social, emotional, and cognitive domains.

Attachment research points to the importance of the parent-child relationship in shaping children's interactions with other children, their sense of security about exploring the world, their resilience to stress, their ability to balance their emotions, their capacity to have a coherent story that makes sense of their lives, and their ability to create meaningful interpersonal relationships in the future. Attachment lays a foundation for how a child comes to approach the world, and a healthy attachment in the early years provides a secure base from which children can learn about themselves and others.

## ATTACHMENT, GENES, AND DEVELOPMENT

An individual's personality develops from a transaction among a child's innate, constitutional temperamental characteristics (such as sensitivity, outgoingness, moodiness) and the experiences that the child encounters as he or she develops within the family and with peers. The genes that children inherit have a large impact on their development, influencing the inborn characteristics of their nervous systems and shaping how people respond to them. Genetic factors can directly shape the way the brain functions and, in this way, have a direct impact on how we behave. These factors are both the genes themselves and the "epigenetic" molecules on the chromosomes that control when, which, and how particular genes will be activated and create the proteins that shape neural structure. Because these epigenetic factors are shaped directly by experience, the activation of our genetic machinery is influenced by what we experience. Experiences directly shape children's development and can influence the specific activation of genes and the sculpting of the connections that make up the structure of the brain. As you can see, the nature-versus-nurture "controversy" is misleading because nature (genes and their regulation) requires nurture (experience) for a child's optimal development. It is not one or the other. Genes and experience interact with each other to shape who we are.

So what can we do? The great news is that you can shape neural growth and function, and even epigenetic regulation, by the experiences you provide! And so here we will continue to dive deep into the specifics of what you can DO to optimize development.

Attachment refers to one very important aspect of the experiential forces that shape a child's development. The human infant is one of the most immature of offspring, having a brain that is quite underdeveloped compared to how complex it will become as the child grows. Also, we as human beings are exquisitely social: our brains are structured to be in relationship with other people in a way that shapes how the brain functions and develops. For these reasons, attachment experiences are a central factor in shaping our development.

Some people worry that the findings of attachment research indicate that our early years create our destiny. In fact, the research shows that relationships with parents can change and as they do the child's attachment changes. This means that it's never too late to create positive change in a child's life. Studies also demonstrate that a nurturing relationship with someone other than a parent in which the child feels understood and safe provides an important source of resilience, a seed in the child's mind that can be developed later on as the child grows. Relationships with relatives, teachers, childcare providers, and counselors can provide an important source of connection for the growing child. These relationships don't replace a secure attachment with a primary caregiver, but they are a source of strength for the child's developing mind.

## SECURE ATTACHMENTS

Secure attachments are thought to occur when children have consistent, emotionally attuned, contingent communication with their parent or other primary caregiver. Relationships that provide contingency, especially at times of emotional need, offer children repeated experiences of feeling connected, understood, and protected. This way of communicating with our children enables them to feel seen, safe, soothed, and secure. We see their internal world beneath their behaviors; we keep them safe from harm and a sense of threat; we tune in to them and soothe their distress, and these experiences enable them to develop a model of security. Studies suggest this security influences both neural connections and those epigenetic regulatory molecules in the brain that support behavior, resilience, and relational well-being. In many ways, this security empowers them to go from our secure home base—our *safe haven*—and explore the world as they use us as a solid *launching pad*.

One way to understand how attachment is created through communication is by looking at what we call the ABC's of the attachment process: attunement, balance, and coherence.

### TABLE 6. THE ABC'S OF ATTACHMENT

The ABC's of attachment are the developmental sequence of attunement, balance, and coherence.

Attunement—Aligning your own internal state with those of your children. Often accomplished by the contingent sharing of non-verbal signals.

Balance—Your children attain balance of their body, emotions, and states of mind through attunement with you.

Coherence—The sense of integration that is acquired by your children through your relationship with them in which they are able to come to feel both internally integrated and interpersonally connected to others.

When a parent's initial response is to be *attuned* to his child, the child feels understood and connected to the parent. Attuned communications give the child the ability to achieve an internal sense of *balance* and supports her in regulating her bodily states and later her emotions and states of mind with flexibility and equilibrium. These experiences of attuned connections and the balance they facilitate enable the child to achieve a sense of *coherence* within her own mind.

Attachment is an inborn system of the brain that evolved to keep the child safe. It enables the child to (1) *seek proximity* to the parent; (2) go to the parent at times of distress for comforting as a source of a *safe haven;* and (3) internalize the relationship with the parent as an internal model of a *secure base.* This sense of security is built upon repeated experiences of being contingently connected with the attachment figure. The impact of these experiences is to provide children with an internal sense of well-being that enables them to go out into the world to explore and make new connections with others.

## INSECURE ATTACHMENTS: AVOIDANCE AND AMBIVALENCE

Parents are not always able to give their child experiences of connection and security so that the child develops a secure attachment. If the ABC's of attachment are not achieved with enough regularity, then the proximity seeking, safe haven, and secure base experiences do not occur optimally. The resulting insecure attachment is carried forward as internal processes in the child that directly influence how the child interacts with others in the future.

Insecure attachments come in several forms and arise from repeated experiences of nonattuned, noncontingent communication. When a parent is repeatedly unavailable and rejecting of the child, a child may become *avoidantly attached,* meaning that the child adapts by avoiding closeness and emotional connection to the parent. Parent-child pairs often have an emotionally barren quality in the tone of their communication. These situations often involve parents who themselves grew up in an emotional desert and have not made sense of that difficult experience and the ways they had to adapt when their attachment needs went unmet.

An *ambivalently attached* child experiences the parent's communication as inconsistent and at times intrusive. The child cannot depend upon the parent for attunement and connection. When children experience inconsistent availability and unreliable communication from their parents they develop a sense of anxiety and uncertainty about whether they can depend upon the parents. They are not sure what to expect. This ambivalence creates a feeling of insecurity in the parent-child relationship, and continues forward in the child's interaction with the larger social world.

In both avoidantly and ambivalently insecure attachments, the children have developed an organized approach to their relationship with their parent. They want to make sense out of their experiences. Children do the best they can to adapt to their worlds. The tenacity of these adaptations can be seen in the ways that children re-create these forms of relationships. How we come to adapt to our primary attachments in our families of origin organizes our minds to approach social relationships in a particular manner that is then applied outside the family. This use of old adaptations in new situations with teachers, friends, and later with romantic partners re-creates yet another experience that reinforces our old patterns of adaptation. We come to deeply believe, for example, that the world is an emotionally barren place (avoidance) or an emotionally unreliable one filled with uncertainty (ambivalence).

## INSECURE ATTACHMENTS: DISORGANIZATION

When children's attachment needs are unmet and their parent's behavior is a source of disorientation or terror, they may develop a *disorganized attachment*. Children with disorganized attachment have repeated experiences of communication in which the parent's behavior is overwhelming, frightening, and chaotic. When the parent is the source of alarm and confusion, children are in a biological paradox. The biological system of attachment is constructed to motivate the child to seek proximity, to get close to a parent at a time of distress in order to be soothed and protected. But in this situation the child is "stuck" because there is an impulse to turn toward the very source of the terror from which he or she is attempting to escape. This is what the attachment researchers Mary Main and Erik Hesse have called "fright without solution." It is an unsolvable dilemma for the child who can find no way to make sense of the situation or develop an organized adaptation. The only possible response of the attachment system is to become disorganized and chaotic.

High rates of disorganized attachment are seen in children who are abused by their parents. Abuse is incompatible with parents' providing children with a sense of security. It fractures the relationship between child and parent and creates an impossible situation for the child's mind by fragmenting a sense of self. Parental abuse has been actually shown to damage the areas of the child's growing brain that enable neural integration. For children with disorganized attachment, impaired neural integration may be one mechanism that leads a child to have difficulty with regulating emotions, trouble in social communication, difficulties with academic reasoning tasks, a tendency toward interpersonal violence, and a predisposition to dissociation— a process in which normally integrated cognition becomes fragmented.

Disorganized attachment is also found in families where, even though physical abuse is absent, the child has repeated experiences in which the parent's behavior is very frightening or in other ways disorienting to the child. Parents who repeatedly rage at their children or become intoxicated may create a state of alarm that leads to disorganized attachment. There is no solution to the paradox that your parent is creating a state of disorientation or terror in you that drives you to seek comfort from the very source of your fear. The disorganizing experiences impair the child's ability to integrate the functions of the mind that enable him to regulate emotions and cope with stress.

Why would parents treat their children like this? Research has demonstrated that parents whose unresolved trauma or loss experiences have not been worked through have a high likelihood of acting out behaviors that terrify their children and create a disorganized attachment for their offspring. Having a history of trauma or loss does not by itself predispose you to having a child with disorganization. It is the lack of resolution that is the essential risk factor. It is never too late to move toward making sense of your experiences and healing your past. Not only you but also your child will benefit.

## DIFFERENT ATTACHMENTS—DIFFERENT COMMUNICATION PATTERNS

When parents understand how attachment affects development and how their communication and behavior affect their child's ability to have a secure attachment to them they are often motivated to change. Parents can learn to have more contingent communication with their children and purposefully work to build a foundation for the develop-

**TABLE 7. PATTERNS OF ATTACHMENT**

| CATEGORY OF ATTACHMENT | PARENTAL INTERACTIVE PATTERN |
| --- | --- |
| Secure | Emotionally available, perceptive, responsive |
| Insecure—avoidant | Emotionally unavailable, imperceptive, unresponsive; and rejecting |
| Insecure—ambivalent | Inconsistently available, perceptive, and responsive; and intrusive |
| Insecure—disorganized | Frightening, frightened, disorienting, alarming |

ment of a healthy parent-child relationship. Reflecting on the categories of attachment may be helpful.

Here are some examples of parent-child interactions that illustrate the attachment categories described above. In each situation a father is taking care of his four-month-old daughter.

*Secure attachment.* The baby is hungry and starts to cry. Her father hears her crying, puts down his newspaper, and goes to her playpen to see what might be causing her distress. He picks her up tenderly, looks into her eyes, and says, "What's wrong, my little darling? Do you want Daddy to play with you? Oh, I know, I bet you're hungry. Is that what you're trying to tell me?" He takes her with him to the kitchen and prepares her bottle, talking to her and telling her that it's almost ready and she'll eat soon. He sits down and, cradling her in his arms, feeds her the bottle. His daughter looks into his face, satisfied by the warm milk and warm interaction with her father. She feels good. Her signals of distress were perceived by her father and he was able to make sense of what they meant and responded to her in a timely and effective manner.

The baby learns from this event, and the many repeated similar contingent connections she experiences with her father, that what she feels internally can be known, respected, and responded to by her father. She feels felt, that someone important in her life knows her. She is able to impact her world with success: "If I communicate, the world will be able to provide me with a way to get my needs met." A secure attachment is developing.

*Avoidant attachment.* A child who develops an avoidant attachment has a different set of experiences with her father. When she cries from her playpen, her father doesn't notice at first. When her crying becomes more insistent, he looks up from his newspaper, but returns to finish the article before going to check on his daughter. He's feeling irritated at the interruption, looks at her in the playpen, and says, "Hey, what's all the fuss?" Thinking she might need her diaper changed, he puts her on the changing table, changes her in silence, and then puts her back in the playpen and returns to the paper. She continues to cry, so he decides that she may need a nap and he places her in her crib. She continues to cry, so he gets her blanket and her pacifier hoping they will calm her, and he closes the door, thinking that she will just take a little time to settle down. She doesn't and it's now forty-five minutes since she initially started to communicate her need for food. "Maybe she's hungry," he realizes, looking at the clock and noticing that it's been over four hours since she's eaten. He gets a bottle ready and she finally calms down when he sits down to feed her.

In this scenario, the baby has learned that her father doesn't always read her signals well. First he has trouble even hearing her, and then he doesn't understand what she wants. He doesn't seem to pay attention to the subtle cues of her communication. Finally, after she's been in distress for quite a while, he gets it right. Overall, repeated patterns like this teach the child that her father is not very available for meeting her needs or connecting to her.

*Ambivalent attachment.* A third baby experiences yet a different response from her father, which creates an ambivalent attachment. When he hears his daughter crying he sometimes knows just what to do. But other times he acts quite anxious and doesn't feel confident that he has the skill to soothe her crying. He gets up from the table where he was reading and runs over to her with a distressed look on his face and picks her up. He's worried and thoughts about work stress come into his mind. It was particularly difficult last week when his boss told him that he wasn't satisfied with his performance and wanted him to show more assertiveness with his clients. This reminds him that his own father continually doubted his abilities and made humiliating comments around the dinner table in the presence of his mother and two older brothers. His mother's anxiety seemed to worsen when his father would start these frequent criticisms and she never stood up for him. Later, she'd go to his room where he had retreated in tears after a scolding by his father and tell him it wasn't right for him to yell back at his father and that he should learn to control himself. She looked very distraught and her anxiety made him quite nervous and even more uncertain of himself. He swore that he would never treat his children like his parents had treated him—and would certainly never do anything to make his children cry.

Here was his daughter still crying in his arms. "This must be one of those times when she is inconsolable," he says to himself. His worried face and his tense arms do not give his daughter a sense of comfort or security. She is only a baby, and cannot know that his anxiety has nothing to do with her own hunger. He soon figures out that she is hungry and gives her a bottle. Although he takes some pleasure in seeing her happy he continues to worry that she will soon start crying again and he won't be able to figure out how to comfort her.

Repeated experiences of this type with her father will lead to an ambivalent attachment with him. This pattern of attachment essentially says, "I'm not certain whether my father will be able to meet my needs, at least in any reliable way. Sometimes he can, and sometimes he can't. Which will it be this time?" Such anxiety creates a sense of uncertainty that others in general can't be relied upon for connection.

*Disorganized attachment.* A fourth child may find that most of her interactions with her father are similar to one of the patterns of the other three babies' experience, but in moments of intense distress her father acts differently. He feels very uneasy when his daughter cries, so as soon as she starts to cry he puts down his newspaper, jumps up, and goes directly to her playpen, hoping to stop her annoying crying. He picks her up abruptly and in his tense state holds her a little too tightly. She initially responds with relief at his arrival, but the tightness of his arms feels more constraining than comforting. She cries louder because she is now both hungry and uncomfortable. Her father senses her increasing distress, which makes him hold on to her even more tightly. He thinks that she may be hungry and carries his crying daughter to the kitchen where he tries to quickly prepare her bottle. Just as he's finishing, the bottle falls and milk pours out over the floor. Surprised by the sound of the bottle hitting the ground, his daughter cries even louder.

Irritated by his own clumsiness as well as his daughter's relentless crying and frustrated by his inability to soothe his daughter, he becomes unable to cope. He feels helpless. His thoughts begin to fragment; memories of his own early history of being mistreated by an alcoholic mother flood him like a tidal wave. His body becomes even tenser, his heart begins to race, his arms tighten as he prepares for his mother's attack on him as he cowers, crying and screaming, under the kitchen table. He hears the sound of glass breaking as his mother drops her bottle of vodka and it turns into shards of glass on the floor around him. She bends down to grab him, and, kneeling on the broken glass, cuts her legs. Furious, she pulls him by the hair and screams in his face that he should "never do that again!"

His daughter is staring off into space, whimpering. Hearing her cries, he realizes he has been somewhere else, off in a trance, and now calls out her name. The flashback is over and he returns to the present, trying to comfort his daughter. She slowly turns back toward him with a vacant look on her face. After a few moments, she looks more fully present. He gets another bottle of milk and sits down to feed her. As she drinks she stares into her father's face, and then looks away at the kitchen floor. He too is shaken by the experience and is only half present. Neither can make any sense of what just happened and each disappears into the disorganization that has developed in his and her mind.

Repeated experiences of her father's entering a trancelike state when she becomes distressed will have a profound impact on the development of his daughter's ability to tolerate and regulate her own intense emotions. These experiences with her father teach her that intense emotion is disorganizing. His overly emotional state creates a disconnection that keeps her from making sense of her own internal and interpersonal world. His entry into a preoccupation with his own unresolved traumas leaves her alone at a time when she desperately needs connection. His nonverbal signals, such as holding her too tightly and the frightened look on his face, create further distress. These experiences are disorienting because they create a sense of fear in his daughter that parallels the fear her father feels of his own mother as well as his fear of being unable to soothe his daughter's distress. These kinds of interactions create a disorganized attachment, and dealing with intense emotion, both her own and others', may be difficult for her in the future. Interpersonal relationships may feel unreliable, and dealing with stress may be particularly challenging in the face of her emerging response of dissociation.

## PATTERNS OF ADAPTATION

Each kind of attachment relationship creates a set of experiences to which the child must respond. For securely attached children, the adaptations will very likely be flexible and promote a sense of well-being. For the organized forms of insecure attachment, avoidance and ambivalence, the adaptations may be less flexible. The biological paradox of disorganized attachment experiences creates a response in the child that lacks organization and does not promote flexibility or enhance the capacity of the child to thrive.

Over time, with repeated experiences, these patterns become incorporated as a characteristic way of being with that parent. These "ways of being" are adaptations or patterns of regulating emotion and intimacy that help to organize both the internal processes of the child's mind and close relationships with others. As children develop, these patterns of response continue to influence their relationships and can become more problematic as they establish their own families.

These attachment categories are a measure of the child's relationship experiences with a specific parent or caregiver. Because the experiences with the parents may be different, a child can be securely attached to one parent but insecurely attached to the other. Attachment appears to have the potential to change over the life span, so if the parent-child relationship changes over time, so can the child's attachment.

Learning to attune, connect, and communicate with your child in the early years creates a secure attachment and lays a foundation for healthy growth and development. A sense of "we" is created that is a living, vital entity. With a secure attachment a child experiences being connected with her parents in a way that enhances her sense of security that gives her a feeling of belonging in the world.

### INSIDE-OUT EXERCISES

1. Consider the three basic attachment elements of proximity seeking, safe haven, and secure base. How do you respond when your children want to be close to you? What about when they are upset and need comforting? Do you think that they are internalizing their relationship with you as a secure base? As your children grow, how do you support their exploration of the world away from you? What might you do to improve your relationship with your children to enhance their security of attachment to you and support their connection and autonomy?

2. Consider ways in which the ABC's of attachment—attunement, balance, and coherence—are a part of your relationship with your child. In what ways are the ABC's comfortable to you—and how are they challenging for you in raising your child?

3. Reflect upon the four categories of attachment: secure, avoidant, ambivalent, and disorganized.

**Genes, Brain Development, and Experience**

The development of the structure and function of the brain is shaped by the interaction of genes and experience. Here are some basic concepts that help clarify some issues about the interrelationships of genes, brain development, and experience.

During gestation, the numerous genes in the nucleus of each cell become expressed and the genes determine what proteins become produced and when and how to shape the body's structure. In utero brain development enables neurons to grow and move to their proper locations in the skull and begin to set up the interconnections that create the circuitry of this complex organ of the nervous system.

When the baby is born, much of the basic brain architecture is already set up, but the interconnections among the neurons are relatively immature compared to their development in the years ahead. During the first three years of life, a massive increase in the connections among neurons creates a complex set of circuitry for the young child. Genetic information influences how neurons become connected during this time, shaping the timing and nature of the emerging circuitry of the brain. As implicit memory exists at this time, we know that these synaptic connections will also be shaped by experience.

One helpful distinction is that of "experience-expectant" versus "experience-dependent" brain development. In expectant development, genetic information determines the growth of neural connections which in turn need to be maintained by exposure to minimal amounts of "expectable" stimulation from the environment. For example, our visual systems need for our eyes to be exposed to light in order to maintain themselves. Without such exposure, the circuits that have been established and are ready to grow and develop will stop their growth and potentially wither away and die. This is an example of "use-it-or-lose-it" brain growth.

In contrast, the experience-dependent growth is thought to involve the connection of neural fibers whose growth is initiated by experience itself. Novel experiences such as interactions with a particular caregiver, putting your feet in a pool of cool water, swinging in the park, or being hugged by your father—all create new synaptic connections in this experience-dependent way.

Some researchers don't make this clear a distinction, pointing instead to the overall nature of brain growth as involving genetically determined overproduction of synapses, "activity-dependent" synaptic growth, and the shaping of brain structure by other means as well, such as the growth of myelin sheaths (which increases the speed of the neural conduction of signals) and blood supply. In this view, neural structure is shaped by numerous factors, including genes and experience, that determine increased growth or destruction of existing neural connections in a process called pruning.

The synaptic density—the number of synaptic connections—in the brain remains high throughout the preschool and elementary school years but a pruning process occurs during adolescence: a naturally occurring destruction of existing neuronal connections which appears to carve specific circuits out of the previously larger mass of interconnected neurons. Pruning is a normal part of development, and even occurs during the early years of brain development. The extent of pruning and specific circuits pruned are determined by experiences and genes, and can be intensified by excessive amounts of stress (caused by the release of high amounts of the stress hormone cortisol for sustained periods of time).

Investigations into the changes occurring during adolescence explore how this natural and extensive process of pruning may lead to reorganizations in the adolescent's brain that help to explain some of the behavioral and emotional experiences of this time. Such restructuring of brain structure produces significant alterations in brain functioning. This process may unveil previously hidden vulnerable aspects of a child's brain architecture created by either genetic inheritance or early stressful experiences that may now become expressed as behavioral and emotional disorders, not merely reorganizations in function, especially with excessive stress-produced pruning.

The important concept that genes and experience interact in shaping development is illustrated in studies by the primate researcher Stephen Suomi and colleagues on rhesus monkeys. These have demonstrated that infant monkeys raised in the absence of maternal care ("peer-raised" monkeys) will show abnormal behaviors, especially if they have a certain variant of a certain gene. When mothers are available to monkeys with this gene, there is a kind of "maternal buffer" such that the gene is not expressed and the behavioral outcome is good. When the absence of maternal care removes this buffer, that gene becomes activated and the ensuing negative effects, which impact serotonin metabolism, become expressed as abnormal regulatory and social behaviors.

The principal message from this and similar studies is that nurturing experiences directly affect how—and even whether—genes become expressed. In the presence of an unhealthy genetic variant, lack of proper nurture can lead to its activation. Further studies reveal that alterations not in the genes themselves but in the molecules that control gene expression are shaped by early experience. These "epigenetic" regulatory molecules help determine which, when, and how genes are expressed and proteins are produced that alter neural connections and, therefore, how the brain functions. Other factors play an important role, too. Studies reveal that we each have different "genetic variants" or "alleles" for different functions, such as how the neurotransmitters in our brains function. If we have a certain variant of dopamine metabolism, for example, we may have more intense reactions to maltreatment than if we are born with another variant. Likewise, there are variants for serotonin or for oxytocin function, an important chemical that influences how sensitive we are to social signals and how we bond with other people. The inborn origin of these variants in our lives can play an important role in how we respond to our close relationships and the ways in which we develop in response to relational experiences. Clearly, the "nature versus nurture" debate needs to be reconfigured to embrace the view that genes *and* experience interact to shape ongoing development.

For parents, an important message from these studies is that the brain is genetically designed to ready itself for normal, healthy development. While we can't change our genes, we can intentionally shape our ways of interacting that influence how those genes are expressed—their epigenetic regulation—and how our children develop. Many neural connections are made and the child is open to receiving love and connection. Parents do not need to be preoccupied with providing excessive sensory stimulation or worried about how to get every neuron to connect to its proper partner! Rather than excessive sensory bombardment, the brain appears to need reciprocal interactions with caregivers in order to grow well. Such nurturing offers the growing brain the shared interactive regulation that is one of the main outcomes of secure attachment relationships.

### Attachment, Experience, and Development

The fields of attachment research and of neuroscience have historically been independent of each other when it comes to studying human beings. Studies of mammals such as primates (monkeys, chimps) and rats have investigated mother-infant experiences and their effects. When we put the independent findings of developmental and cognitive neuroscience together with the nonhuman and human studies of attachment, an extremely exciting vision of the "larger reality" of human development begins to appear. It suggests that the attachment of the infant and older child to the caregiver provides a set of experiences that entail the ABC's of attachment: attunement, balance, and coherence.

Attunement is the parent's aligning of her internal state with that of the child. This process often involves the sharing and coordination of nonverbal signals (eye contact, facial expression, tone of voice, gestures and touch, bodily posture, timing, and intensity of response). Such a nonverbal resonance likely involves a connecting process between the right hemispheres as they mediate nonverbal signals in both people.

Balance is the regulation that the physical presence and the attuned communication of the parent provide to the immature and growing brain of the child. In this way, attuned and contingent communication provides the external connecting process that enables the child to achieve internal states of balance. This balance involves the regulation of processes such as sleep-wake cycles, responses to stress, heart rate, digestion, and respiration. The developmental neuroscientist Myron Hofer has called these "hidden regulators" through which the maternal presence enables the offspring's basic physiological balance to be achieved. Prolonged maternal absence stresses the very young animal by removing these hidden regulators and exposing a very young brain to debilitating amounts of unregulated stress in her absence. Such repeated stressors have been shown to have prolonged negative effects on the animal's ability to have balanced physiological regulation in the future.

Coherence is the outcome of successful parent-mediated balance in which the brain becomes adaptive, stable, and flexible to adjust to changing environmental demands. A well-integrated, organized brain creates a coherent, adaptive mind. This view is supported by attachment research that suggests that secure attachments filled with integrative communication in which differences are honored and compassionate communication cultivated promote a coherent mind, whereas insecure attachments generate various forms of incoherence.

An incoherent mind is seen in the extremes of child abuse and neglect. Research on abused and neglected children has revealed the devastating effects of maltreatment on the child's growing brain: smaller overall brain size, decreased growth of the corpus callosum, which connects the right and left sides of the brain, negative impacts on the integrative prefrontal and hippocampal regions, and impaired growth of the GABA (gamma-aminobutyric acid, an inhibitory neurotransmitter) fibers from the cerebellum that normally serve to calm the excitable emotional limbic structures—all are seen in brain imaging studies of abused and neglected children. The probable source of these difficulties is the excessive amount of stress hormone re-leased during these traumatic events, which is toxic to neurons, im-pairing their growth and killing existing cells. We know that these effects on the brain occur, but we don't know whether positive expe-riences in the future can help to overcome the neurological effects. We know that people can recover from abuse through nurturing rela-tionships, but we do not know whether brain damage is repaired or whether alternative circuits are developed in the healing process.

There is a simple principle that may reveal the important link between our relationships with our children and the growth of their brains. When communication is integrative, the integrative fibers of the brain grow well. Since neural integration in the brain is the basis for healthy regulation—of attention, emotion, and behavior—then we can see the possible truth beneath this statement: health emerges from integration within us, and between us. Your central principle, then, is to offer integrative experiences so that the integrative fibers in your child (and in yourself) that support healthy forms of regulation of all sorts can grow well!

These scientific findings taken together support the assertion that what parents do matters. There is no question that nature needs nurture: the nurturing ways parents interact with their children can have a positive effect on the developing mind by laying the ground-work for healthy brain development, which can serve as a foundation for patterns of secure attachment across the generations.

■ ■ ■

### The World of Attachment Research

Research on attachment has a long and rich history. John Bowlby (1907–1990), an English physician and psychoanalyst, thought that the actual experiences that children had with their parents played an important role in providing them with an internal sense of well-being that he called a "secure base." Bowlby's ideas influenced the ways that children were treated in hospitals and orphanages so that instead of the prior practice of having rotating caregivers (to avoid the pain of separation when the children had to leave the institutions), "primary caregivers" were assigned to children, who could get to know them and develop an attachment to them. Children in these changed institutional settings went from dying to thriving.

Mary Ainsworth (1913–1999), a research psychologist from Canada, worked with Bowlby and developed a research approach to testing these ideas. Her groundbreaking work established the three categories of attachment of secure, insecure avoidant, and insecure ambivalent. She designed a measure called the "infant–strange situation" which is the gold-standard method for testing a one-year-old's attachment to a given caregiver. In the laboratory setting, an infant is brought into a room with toys, a stranger, and a two-way mirror so the event can be filmed. First the infant is left alone with the stranger; then the parent returns; then, both the parent and the stranger leave for up to three minutes; then the parent returns. The interaction during the parent's return is assessed. In this separation scenario the most useful data is the reunion behavior of the baby when the parent returns.

Securely attached infants may be upset during separation, but during the reunion they seek proximity, are quickly soothed, and readily return to play and exploration. The avoidantly attached infants seem to act as if the parent never left the room, continuing about their business of playing with the toys and ignoring the parent's return. Their external behavior appears to say, "I haven't found our interactions useful in the past, what good does it do for me to go to you now?" Measurement of their physiological stress response, however, reveals that they notice quite well the parent's return. The ambivalently attached children quickly seek proximity with the returned parent but they are not easily soothed and they do not readily return to playing. They cling to the parent, looking as if they are filled with uncertainty about the parent's ability to soothe and protect.

Ainsworth's studies demonstrated that the data collected during direct observations of the parent-child relationship during the first year of life could allow researchers to predict with high degrees of certainty what the "infant–strange situation" result would be. In general, parents who were sensitive to their child's signals had babies who were securely attached to them. Parents who were neglectful or rejecting tended to have children who were avoidantly attached to them. Parents who were inconsistently available and who may have been intrusive generally had children with an ambivalent attachment to them.

* * *

### Deepening Our Understanding of Attachment

After initially finding correlations between parent-child patterns of communication and an infant's behavior at one year of life in the "infant–strange situation" studies, a number of researchers began to follow these children in long-term, or longitudinal, studies. Alan Sroufe and his colleagues in Minnesota have carried out one of the longest longitudinal studies of this kind and have found some fascinating ways in which early attachment profiles can be used to predict later child outcomes. Their ingenious studies have taken them into school classrooms and summer-camp settings in pursuit of data on how these kids interact with others, how their peers perceive them, and how they interact with other adults outside the home. Though the pattern of attachment from early on can change if significant relationships change in the child's life, stable attachment relationships are correlated with a number of important outcomes: securely attached babies grow into kids with leadership ability, avoidantly attached babies later are shunned by their peers, ambivalently attached babies become anxious and uncertain kids, and disorganizedly attached babies when older have the most difficulty getting along with others and balancing their own emotions.

Mary Main and her colleagues at the University of California, Berkeley, have made major contributions to both our knowledge foundation and our ways of thinking about attachment. They defined the disorganized category of attachment. In the "infant–strange situation" test, these babies exhibited chaotic and disorienting behavior during the reunion with the parent. Main and Erik Hesse, her colleague and husband, have proposed that the parent's frightening, fearful, or disorienting behaviors create a state of alarm, or "fright without solution" which is the experiential source of this unusual and disturbing disorganization response. Main and her colleagues have also moved our understanding of attachment into the realm of the adults' "state of mind with respect to attachment," which provides us with deep insights into why parents behave in such different ways with their children, which results in these differing categories of attachment. In fact, Main and the numerous studies she has inspired have demonstrated that the adult's state of mind with respect to attachment is the most robust predictor of a child's attachment. More on this in the next chapter!

## Digging Deeper

M. Ainsworth et al. *Patterns of Attachment: A Psychological Study of the Strange Situation*. Hillsdale, N.J.: Erlbaum, 1978.

J. Bick and M. Dozier. "Helping Foster Parents Change: The Role of Parental State of Mind." In H. Steele and M. Steele, eds., *Clinical Applications of the Adult Attachment Interview*, pp. 452–470. New York: Guilford Press, 2008.

Y. A. E. Bigelow, K. MacLean, J. Proctor, T. Myatt, R. Gillis, and M. Power. "Maternal Sensitivity Throughout Infancy: Continuity and Relation to Attachment Security." *Infant Behavior and Development* 33, no. 1 (2010): 50–60.

J. Bowlby. *A Secure Base: Parent-Child Attachment and Healthy Human Development*. New York: Basic Books, 1988.

T. B. Brazelton and S. I. Greenspan. *The Irreducible Needs of Children*. Cambridge, Mass.: Perseus Publishing, 2000.

J. Cassidy and P. R. Shaver, eds. *Handbook of Attachment: Theory, Research, and Clinical Applications*, second edition. New York: Guilford Press, 2008.

M. Dozier, K. C. Stovall, K. E. Albus, and B. Bates. "Attachment for Infants in Foster Care: The Role of Caregiver State of Mind." *Child Development* 72 (2001): 1467–1477.

M. Dozier, O. Lindhiem, E. Lewis, J. Bick, K. Bernard, and E. Peloso. "Effects of a Foster Parent Training Program on Young Children's Attachment Behaviors: Preliminary Evidence from a Randomized Clinical Trial." *Child and Adolescent Social Work Journal* 26, no. 4 (2009): 321–332.

W. T. Greenough and J. E. Black. "Induction of Brain Structure by Experience: Substrates for Cognitive Development." In M. R. Gunnar and C. A. Nelson, eds., *Minnesota Symposia on Child Psychology* 24:155–200. Hillsdale, N.J.: Erlbaum, 1992.

M. Hofer. "On the Nature and Consequences of Early Loss." *Psychosomatic Medicine* 58 (1996): 570–581.

A. Holtmaat and K. Svoboda. "Experience-Dependent Structural Synaptic Plasticity in the Mammalian Brain." *Nature Reviews Neuroscience* 10, no. 9 (2009): 647–658.

A. F. Lieberman and P. Van Horn. *Psychotherapy with Infants and Young Children: Repairing the Effects of Stress and Trauma on Early Attachment*. New York: Guilford Press, 2008.

M. Main. "Attachment: Overview, with Implications for Clinical Work." In S. Goldberg, R. Muir, and J. Kerr, eds., *Attachment Theory: Social, Developmental and Clinical Perspectives*, pp. 407–474. Hillsdale, N.J.: Analytic Press, 1995.

M. J. Meaney. "Epigenetics and the Biological Definition of Gene X Environment Interactions." *Child Development* 81, no. 1 (2010): 41–79.

M. Rutter. "Attachment Reconsidered: Conceptual Considerations, Empirical Findings, and Policy Implications." In J. P. Shonkoff and S. J. Meisels, eds., *Handbook of Early Childhood Intervention,* second edition, pp. 651–682. New York: Cambridge University Press, 2000.

M. Rutter, J. Kreppner, and E. Sonuga-Barke. "Emanuel Miller Lecture: Attachment Insecurity, Disinhibited Attachment, and Attachment Disorders: Where Do Research Findings Leave the Concepts?" *Journal of Child Psychology and Psychiatry* 50, no. 5 (2009): 529–543.

J. R. Schore and A. N. Schore. "Modern Attachment Theory: The Central Role of Affect Regulation in Development and Treatment." *Clinical Social Work Journal* 36, no. 1 (2008): 9–20.

J. P. Shonkoff and D. A. Phillips. *From Neurons to Neighborhoods: The Science of Early Childhood Development.* Washington, D.C.: National Academy Press, 2000.

D. J. Siegel. *The Developing Mind, Second Edition: How Relationships and the Brain Interact to Shape Who We Are.* New York: Guilford Press, 2012. Chapter 3.

L. A. Sroufe. "The Concept of Development in Developmental Psychopathology." *Child Development Perspectives* 3, no. 3 (2009): 178–183.

L. A. Sroufe, E. A. Carlson, A. K. Levy, and B. Egeland. "Implications of Attachment Theory for Developmental Psychopathology." *Development and Psychopathology* 11 (1999): 1–13.

L. A. Sroufe, B. Egeland, E. Carlson, and A. Collins. *The Development of the Person.* New York: Guilford Press, 2005.

L. A. Sroufe and D. J. Siegel. "The Verdict Is In: The Case for Attachment Theory." *Psychotherapy Networker,* 2011. See www.psychotherapynetworker.org/magazine /recentissues/1271=the=verdict=is=in.

S. Suomi. "A Biobehavioral Perspective on Developmental Psychopathology: Excessive Aggression and Serotonergic Dysfunction in Monkeys." In A. J. Sameroff, M. Lewis, and S. Miller, eds., *Handbook of Developmental Psychopathology,* second edition. New York: Plenum Press, 2000.

L. M. Supkoff, J. Puig, and L. A. Sroufe. "Situating Resilience in Developmental Context." In M. Ungar, ed., *The Social Ecology of Resilience,* pp. 127–142. Springer: New York, 2012.

M. Teicher. "The Neurobiology of Child Abuse." *Scientific American* (March 2002): 68–75.

R. A. Thompson. "Early Attachment and Later Development." In J. Cassidy and P. R. Shaver, eds., *Handbook of Attachment: Theory, Research and Clinical Applications,* second edition, pp. 265–286. New York: Guilford Press, 2008.

M. H. van Ijzendoorn and M. J. Bakermans-Kranenburg. "DRD4 7-Repeat Polymorphism Moderates the Association Between Maternal Unresolved Loss or Trauma and Infant Disorganization." *Attachment and Human Development* 8, no. 4 (2006): 291–307.

M. H. van Ijzendoorn, K. Caspers, M. J. Bakermans-Kranenburg, S. R. H. Beach, and R. Philibert. "Methylation Matters: Interaction Between Methylation Density and Serotonin Transporter Genotype Predicts Unresolved Loss or Trauma." *Biological Psychiatry* 68, no. 5 (2010): 405–407.

M. H. van Ijzendoorn and A. Sagi. "Cross-Cultural Patterns of Attachment: Universal and Contextual Dimensions." In J. Cassidy and P. R. Shaver, eds., *Handbook of Attachment: Theory, Research and Clinical Applications,* second edition, pp. 713–734. New York: Guilford Press, 2008.

N. S. Weinfield, G. J. L. Whaley, and B. Egeland. "Continuity, Discontinuity, and Coherence in Attachment from Infancy to Late Adolescence: Sequelae of Organization and Disorganization." *Attachment and Human Development* 6, no. 1 (2004): 73–97.

C. H. Zeanah, J. A. Larrieu, S. S. Heller, and J. Valliere. "Infant-Parent Relationship Assessment." In C. Zeanah, ed., *Handbook of Infant Mental Health,* third edition, pp. 266–280. New York: Guilford Press, 2009.

# CHAPTER 6

# How We Make Sense of Our Lives: Adult Attachment

## INTRODUCTION

Children's lives are in part a chapter in their parent's life story. Each generation is influenced by the preceding generations and influences future generations. Even though our own parents did the best they could, given the circumstances of their own lives, we may not have had the early experiences that we would wish to pass on to our own children.

The positive connections we have had with others in or outside of our families serve as a core of resilience that may have helped us to weather the storm of difficult times in the past. Fortunately, even those of us who had quite difficult childhoods often have had some positive relationships during those years that can offer a seed of strength to help us overcome early adversity.

We are not destined to repeat the patterns of our parents or of our past. Making sense of our lives enables us to build on positive experiences as we move beyond the limitations of our past and create a new way of living for ourselves and for our children. Making sense of our own lives can help us to provide our children with relationships that promote their sense of well-being, give them tools for building an internal sense of security and resilience, and offer them interpersonal skills that enable them to make meaningful, compassionate connections in the future.

How we have come to make sense of our lives, how we tell a coherent story of our early life experiences, is the best predictor of how our children will become attached to us. Adults who have made sense of their lives have an adult security of attachment and are likely to have children who are securely attached to them. Enabling our children to build a secure attachment lays a foundation for their future healthy development.

## PARENTS MAKING SENSE OF THEIR LIVES

Reflecting on your childhood experiences can help you make sense of your life. Since the events of your childhood can't be altered, why is such reflection helpful? A deeper self-understanding changes who you are. Making sense of your life enables you to understand others more fully and gives you the possibility of choosing your behaviors and opening your mind to a fuller range of experiences. The changes that come with self-understanding enable you to have a way of being, a way of communicating with your children, that promotes their security of attachment.

Our life stories can evolve as we grow throughout the life span. The capacity to integrate the past, present, and future allows us to move into more coherent levels of self-knowledge. This increased coherence of our life story is associated with movement toward an adult security of attachment. Changing attachment status as we develop is quite possible. Studies have shown that individuals can move from what was an insecure childhood attachment to a secure adult attachment status. These studies examine the finding of an "earned-security" status, one that is important for our understanding of coherent functioning and the possibilities for change. Individuals with earned security may have had troubled relationships with their parents during their own childhoods, but they have come to make sense of their childhood experiences and their impact on their development as adults. Relationships, both personal and therapeutic, appear to be able to help an individual develop from an incoherent (insecure) to a more coherent (secure) functioning of the mind. Such growth is carried out through relationships that help an individual to heal old wounds and transform defensive approaches to intimacy.

Here is how one parent with a secure attachment and a securely attached four-year-old son tells a coherent story about her childhood.

"My mother and father were very caring, but my father had a problem with manic-depressive illness. It really made growing up at my house unpredictable for my sisters and me. It helped that my mother knew how scared I was of my father's moods. She was very tuned in to me and did her best to help me feel safe. It was very frightening, although at the time I thought being scared was normal. But in thinking about it now I realize how much my father's unpredictability shaped so much of my life as a child, and even into my twenties.

"It wasn't until he was treated effectively after my son was born that I could get any perspective on what had happened. Initially it was really hard for me to deal with my son when he got upset. I got scared all over again at somebody out of control. I had to try to figure out why I had such a short fuse. I really had to work on myself so I could be a better parent. I'm not so reactive and my son and I get along great. My relationship with my dad is even getting better now. Our relationship had been very stormy, but now I think it's pretty good. He is not as sensitive or open as my mother is, but he's doing the best he can. He's gone through a lot and you've got to respect him for that."

This woman had a difficult time growing up but has made sense of her painful experiences. She recognizes the impact of her childhood relationships, both good and bad, on her development and her role as a parent. Her reflections have an open quality to them as she continues the lifelong process of making sense of her life. It is fortunate for her child that she has earned her security as an adult and is free to nurture him in ways that can promote his own sense of vitality and connection in the world.

Nurturing relationships support our growth by helping us to make sense of our lives and to develop the more reflective, integrated functioning that emerges from secure attachments. There is always hope and possibility for change.

## ADULT ATTACHMENT

Each of us has a state of mind or general stance toward attachment that influences our relationships and is often revealed in how we tell our life stories. These states of mind have been studied and found to be associated with specific patterns of communication and interaction between parent and child, which shape the ways in which the child develops a security or insecurity of attachment.

The attachment researcher Mary Main and her colleagues thought that perhaps some elements of the parents' own history of being a child would be important in determining their behavior toward their own children. They came up with a research instrument, called the Adult Attachment Interview, in which parents are asked about their recollections of their own childhoods. As it turned out, the way parents made

## TABLE 8. ATTACHMENT CATEGORIES AS CHILDREN AND AS ADULTS

| CHILD | ADULT |
| --- | --- |
| Securely attached | Secure (free or autonomous) |
| Avoidantly attached | Dismissing |
| Ambivalently attached | Preoccupied |
| Disorganizedly attached | Unresolved trauma or loss/ disorganized |

sense of their own early life experiences, as revealed in the coherence of their life narrative as told to an interviewer, is the most powerful feature that could predict the child's security of attachment. The table above shows the correlation between child attachment and adult attachment.

Adult attachment can be determined by how parents tell the story of their early life history to another adult. Parents' understanding of themselves is revealed through this adult-to-adult communication, not through how parents explain their early histories to their own children. The way the story is told, not merely the content, reveals characteristics of the parent's state of mind regarding attachment. These narrative patterns are associated with the child's attachment status to that parent as demonstrated in the above table. Long-term studies have further shown that adults' narratives generally corresponded to their own childhood attachment categories assessed decades earlier.

When you read about these categories, it is important not to try to classify yourself rigidly in one particular grouping. It's normal to have some elements of several categories. Children often have different patterns of attachment to the different adults in their lives. We can carry those different "models of attachment" or "states of mind" forward in our lives and activate them in different settings. For example, you may have some history of both avoidance and of ambivalence. If you are in relationship with someone who is emotionally distant, it may evoke in you avoidant ways of responding that come from a dismissing "state of mind with respect to attachment." If later on you are relating to someone who is intrusive and inconsistent, that may evoke your ambivalent model and you may act from a preoccupied state of mind when you are with that person. While research requires one dominant category to be given to a subject in the investigation, our professional experience has shown that while we may have a dominant model that emerged with our primary caregiver, we all may potentially show several other models of attachment activated by relationships with others in various settings. The ideas explored by means of these categories can be useful in deepening the kind of self-understanding that promotes secure attachment in our children when considered in a flexible and supportive manner.


**Secure Adult Attachment**

An autonomous or free state of mind with respect to attachment is usually found in adults with children who are securely attached to them. These adults' narratives are characterized by a valuing of relationships and a flexibility and objectivity when speaking about attachment-related issues. These individuals integrate their past with the present and their anticipated future. These are coherent narratives revealing how the person has made sense of his or her life history. Emerging research that has followed infants into their young adulthood suggests that such narratives develop when the individuals have had secure attachments when they themselves were children.

An earned secure adult attachment is one in which the narrative is coherent, but there were difficult times in childhood attachment.

Earned security reflects how an adult has come to make sense of his or her early life history. The story earlier in this chapter about the woman with a manic-depressive father is an example of how one parent with earned security might describe her experiences.

## Dismissing Adult Attachment

For adults whose early life may have included a predominance of parental emotional unavailability and rejection, a dismissing stance toward attachment may be found. As parents their children's relationships with them are often characterized by avoidant attachments. These parents appear to have little sensitivity to the child's signals. The inner world of such adults seems to function with independence as its hallmark: being disconnected from intimacy and perhaps even from the emotional signals of their own bodies. Their narratives reflect this isolation, and they frequently insist that they do not recall their childhood experiences. Life seems to be lived without a sense that others or the past contribute to the evolving nature of the self. One proposal is that these adults live in a predominantly left mode of processing, especially when interacting with others. In spite of these mechanisms that minimize the importance of relationships, a number of studies suggest that the children and adults in this grouping have bodily reactions that indicate that their nonconscious minds still value the importance of others in their lives. Their behavior and their conscious thoughts appear to have adapted to an emotionally barren home environment by producing an avoidant and dismissing stance toward attachment and intimacy.

In response to questions about her childhood experiences a mother of an avoidantly attached child made the following self-reflective comments:

"My own mother and father lived at home and created the kind of house that is a good one for a child to grow in. We went to many activities and I was given all of the experiences any child would expect from a good home. For discipline we were taught right from wrong and given the right direction for how to succeed in life. I don't remember exactly how they did this, but I do know it was a good childhood overall, a regular childhood in the sense that it was good. That's about it. Yes, it was a good life."

Notice that though this adult is able to have a cohesive, logically consistent story of her life with a general theme that it was a "good life," little detail is offered from personally experienced memories to illustrate that view in an enriched and coherent way. Coherence, making sense, involves a more holistic, visceral process of reflection. For example, the statement "For discipline we were taught right from wrong and given the right direction for how to succeed in life" does not involve an autobiographical sense of the self in time as personally lived. A coherent narrative might have reflections such as, "My mom really tried to teach me right from wrong, though I didn't always listen and she would get mad at me at times. I remember one day I picked all the neighbor's flowers to make a bouquet for her and she told me that she loved the thought, but that it was wrong to take something without permission. I felt so bad when we had to give the flowers back and take them a potted plant as an apology." This mother's actual reflections reveal a stance that minimizes emotional vulnerability and dependence on others. Her shared story has many words that logically state her thoughts in a linear fashion, but little sense that there is an integration of memory, emotion, relatedness, and a process connecting the past to the present and how these might influence the future.

### Preoccupied Adult Attachment
Adults who have experienced inconsistently available, perceptive, and responsive caregiving often have a preoccupied stance toward attach-

ment, one filled with anxiety, uncertainty, and ambivalence. These preoccupied parental states may impair the ability of the parents to reliably perceive their children's signals or interpret their needs effectively. Their children are often ambivalently attached to them. The parents may be overwhelmed by doubts and fears about relying upon others. Their stories are often filled with anecdotes revealing how leftover issues from the past continue to enter the present and steer the narrative flow away from addressing the topic at hand. This intrusive pattern of leftover issues is a direct impairment to mindfulness and may block the capacity for flexibility. The incoherence of these intrusions may be understood as a flooding of the right mode of processing into the left mode's attempt to make logical sense of the past.

One father of an ambivalently attached child made these comments in response to questions about his childhood experiences:

"My childhood growing up? It was something else! We were very close, not too close, but close enough. We used to have lots of fun together, me and my two older brothers. Sometimes they'd get pretty rough, me, too, but it wasn't a problem, though my mother thought it was. Sometimes. Even this weekend, I guess it was Mother's Day, and she thought we were too hard on our kids. I mean, she said I was too rough on my son. But she never said that about my two brothers when we were kids. I mean, she let them tackle me for God's sakes and didn't say a word to them. No. It's always me. But I don't care, it doesn't bother me anymore. Or it could. But I wouldn't let it happen. Should I?"

These responses reveal how themes from the past intruded on this parent's ability to reflect coherently on his life. His focus on the past merges into discussions of a recent weekend, then back to his childhood, and again into his present preoccupations. He is still enmeshed in the issues of his childhood. This leftover baggage will likely interfere with his ability to connect with his own children. For example, if he feels slighted by his son's seeking attention from his wife, he may become flooded with a sense of being treated unfairly, similar to how he experienced his mother's favoritism of his brothers. If he doesn't make sense of and resolve these leftover issues, he will be prone to create emotionally clouded interactions with his own children.

### Unresolved Adult Attachment

Parental unresolved trauma or loss is often associated with the most concerning child attachment category, disorganized attachment. Parents with unresolved trauma appear to enter abrupt shifts in their states of mind that are alarming and disorienting to their children. Examples of such behavior include spacing out when a child is distressed, becoming suddenly enraged and threatening to a child who is excitedly skipping down the hallway singing "too loudly," or hitting a child who asks for another bedtime story.

What is it about unresolved trauma and grief that produces such alarming and disorienting parental behaviors? Unresolved conditions involve a disruption in the flow of information in the mind and in the self's ability to attain emotional balance and maintain connections to others. This impairment is called "dysregulation." Abrupt shifts in the flow of information and energy can occur within a person or between people. Moods may become sullen, emotions abruptly change without warning, and perceptions become colored by sudden shifts in attitude. Responding to change may be difficult and result in inflexibility.

These internal processes directly influence interpersonal interactions: unresolved states produce these abrupt shifts that can create a state of alarm in a child. These are the likely origins of the "fright without solution" experience, as the parent not only disengages from contingent communication but engages in terrifying behaviors. The child experiences the sudden absence of something positive with the emergence of something quite negative.

How are such dysregulated internal processes revealed in an adult's self-reflections? There may be moments when an individual may become disoriented while discussing issues of trauma or loss. This momentary lapse in what may otherwise be a coherent story is thought to reveal some fundamental way in which those topics remain unresolved for that person. A child with a disorganized attachment can become immersed in the chaos that is her parent's internal legacy of a chaotic past.

Here are one mother's reflections on questions as to whether she had felt threatened in her childhood:

"I don't think I ever really felt threatened when I was a kid. I mean, not that it wasn't scary. But it was. My father would come home drunk sometimes, but my mother was the real kicker. She had this way of trying to get you to trust her, but it was his drinking that she'd blame, blame for her being so mean. I mean, she was trying to be nice, but it's as if she became possessed by a demon. Her face would change suddenly and you never knew who to trust. I mean, she looks so strange, kind of furious and frightened in one package, all contorted and her eyes so piercing. She would sometimes cry for days. I can see her crying face now. Not that it was that upsetting, but it was."

When this woman was a child and was exposed to the sudden abrupt shifts in her mother's face when her mother was exceedingly angry or sad, her own mind might have been readying itself to create such intense emotions in the future. Such a process may be due to the mirror neuron system that may create in us an emotional state similar to that which we perceive in others. A parent's disorienting behaviors will create a state of disorganization, and disorganized attachment, in the child. The child, empathically through mirror neurons and directly through her own fright without solution, enters her own state of internal chaos.

Unresolved states can flood the normally smooth flow of internal processing and contingent interpersonal communication. Lack of resolution may be revealed during the telling of an individual's life story, and it also may predispose that person to lose the capacity to be flexible under certain conditions related to the trauma or loss. Their capacity to integrate left and right modes of processing may be quite impaired, which leads to disorganization when reflecting on their lives and discussing unresolved topics. When autobiographical recollections are being reactivated, the left hemisphere may become flooded by the unprocessed images and sensations of terror and betrayal. Rather than having a coherent sense of how the past has impacted on the present, these abrupt, dysregulated processes flood the individual and immerse him or her in the chaos of the past. Trauma-related emotional reactivity, a reactivating set of mirror neurons, and impaired integration across the hemispheres may possibly all contribute to these internal processes that create the incoherence and interpersonal chaos of unresolved trauma and loss.

### REFLECTIONS ON ATTACHMENT

The questions on pages 146–147 can help you reflect on your own childhood experiences. These "Questions for Parental Self-Reflection" are not a research tool such as the Adult Attachment Interview, but have been helpful to many individuals for deepening their self-understanding. Use these questions in a flexible way and consider other issues that may serve as cues to your memory. Feelings and images may come to you in the course of answering these questions. Sometimes we may feel uncertain or ashamed of parts of our memories and begin to edit what we say in order to present a more pristine image of ourselves to others, or to ourselves. Each of us may have ghosts in our closets! There may be patterns of defensive adaptations that have kept us distant from our own emotions and prevented us from being open to the feelings of others.

Initially it may be quite difficult to find words to express our internal images or sensations. This is normal. Recall that words and conscious verbal thoughts emerge from our left hemispheres whereas autobiographical memory, raw emotion, integrated bodily sensation, and imagery are processed nonverbally in our right hemispheres. This situation creates a tension in the translation of the nonverbal into the verbal, especially if the autobiographical, emotional, and visceral memories are overwhelming and unprocessed. Sometimes recollections of painful elements of our emotional memories can make us feel very vulnerable. However, not owning the totality of our histories can actually inhibit the creation of coherency in our lives. Embracing all that we are can be stressful and difficult at first, but it eventually leads to a sense of compassionate self-acceptance and interpersonal connection. Integrating a coherent story involves bringing together the themes from our past with the ongoing story of our lives as we move into the future.

Change happens through a process of trying new ways of relating that can support your journey into deeper levels of self-understanding. In self-reflection, it can be helpful to find an adult whom you trust who can listen to aspects of your evolving journey of discovery. We are all social beings, and our narrative processes are born out of social connections. Self-reflection is deepened when it is shared within our intimate relationships.

1. What was it like growing up? Who was in your family?

2. How did you get along with your parents early in your child-hood? How did the relationship evolve throughout your youth and up until the present time?

3. How did your relationship with your mother and father differ and how were they similar? Are there ways in which you try to be like, or try not to be like, each of your parents?

4. Did you ever feel rejected or threatened by your parents? Were there other experiences you had that felt overwhelming or traumatizing in your life, during childhood or beyond? Do any of these experiences still feel very much alive? Do they continue to influence your life?

5. How did your parents discipline you as a child? What impact did that have on your childhood, and how do you feel it affects your role as a parent now?

6. Do you recall your earliest separations from your parents? What was it like? Did you ever have prolonged separations from your parents?

7. Did anyone significant in your life die during your childhood, or later in your life? What was that like for you at the time, and how does that loss affect you now?

8. How did your parents communicate with you when you were happy and excited? Did they join with you in your enthusiasm? When you were distressed or unhappy as a child, what would happen? Did your father and mother respond differently to you during these emotional times? How?

9. Was there anyone else besides your parents in your childhood who took care of you? What was that relationship like for you? What happened to those individuals? What is it like for you when you let others take care of your child now?

10. If you had difficult times during your childhood, were there positive relationships in or outside of your home that you could depend on during those times? How do you feel those connections benefited you then, and how might they help you now?

11. How have your childhood experiences influenced your relationships with others as an adult? Do you find yourself trying *not* to behave in certain ways because of what happened to you as a child? Do you have patterns of behavior that you'd like to alter but have difficulty changing?

12. What impact do you think your childhood has had on your adult life in general, including the ways in which you think of yourself and the ways you relate to your children? What would you like to change about the way you understand yourself and relate to others?

## PATHWAYS TOWARD GROWTH

As you reflect on your own attachment history, you may find that some dimensions are particularly relevant for understanding the impact of early family experiences on your development as an adult. We have found that the general framework provided by attachment research is useful for deepening self-understanding and for pointing to some pathways toward change. Research demonstrates that growth toward security of attachment is quite possible. Though the movement toward security is often associated with healthy and healing relationships with friends, lovers, teachers, or therapists, beginning with the process of deepening your self-understanding can serve as an invitation to enhance your connections with others. Moving toward security offers an enriched way of life for both you and your children.

### Avoidance and a Dismissing Stance

For those whose histories included a sense of emotional unavailability and a lack of attuned, nurturing parenting, there may have been an adaptation that minimizes the importance of interpersonal relationships and the communication of emotion. This minimizing stance may have been very adaptive to children raised in an emotional desert. Children do the best they can, and reducing dependence on emotionally unavailable caregivers may have been an appropriate and useful adaptation for their survival. As this adaptive response continues, children may have a decreased connection not only to their parents but to other people as well. Although studies reveal that individuals with avoidant attachment have mindsight and can take others' perspectives, their defensive state of mind appears to reduce their motivation to be open to the emotional experiences of others. In addition, there may be a decreased access to and awareness of their own emotions. These avoidant adaptations can be seen as a minimization of right-mode processing in favor of a predominance of left-mode thinking in order to reduce their emotional vulnerability.

With this perspective, one can imagine that the degree of bilateral integration connecting the right and left modes of processing may be quite minimal as well. This may be revealed in how people with this early history do appear to have very underdeveloped life stories. Often they insist that they don't remember the details of their early life experiences. Within relationships, there can be a marked sense of independence that may lead their partners to experience loneliness and emotional distance. The processes that began as "healthy" and necessary adaptations in childhood may become impediments to healthy adult relationships with one's spouse and children.

Approaches toward changing these adaptations are those that promote bilateral integration. Often right-mode processing may be underdeveloped, demonstrated by the possession of minimal mind-sight abilities, diminished self-awareness, and at times a decreased ability to perceive the nonverbal signals of others. Self-reflection may be limited because it is primarily the logical and nonautobiographical left mode that becomes engaged. For this reason, efforts to activate the right mode may be helpful and necessary. Research has shown the presence of heightened physiological reactions in these people during discussion of attachment-related issues, in contrast to their comments minimizing the importance of attachments. This indicates the reality that their minds respond as though relationships are significant, even if behavior and overt attitude do not reveal this emotional valuing of connections. In other words, the inborn attachment system that values relationships is still intact, even though the person's adaptation has required a minimization approach. This view is crucial in considering how to reach out to individuals who may have spent much of their lives automatically minimizing the importance of relationships in order to adapt to past deficits in family life. Those families can be seen as having offered little right-mode stimulation and connection. Finding a way to activate these aspects of mental life can be crucial in liberating the mind's innate drive toward interpersonal connection and internal integration. We have found that activities such as guided imagery that focus on nonverbal signals, increase awareness of sensations in the body, and stimulate the right side of the brain have been quite useful to mobilize underdeveloped right-mode processes.

For someone who is steeped in logic, it is actually helpful to offer the logical explanation described above that an emotionally distant family environment early in their lives might have contributed to their left hemispheres' becoming adaptively dominant. It may also be helpful to point out that recent findings from brain science suggest that new neurons, especially integrative ones, may continue to be able to grow throughout our lives. With this relatively neutral, nonthreatening perspective, the work to open up the gates toward the growth and integration of the other equally important but less well developed mode can begin.

### Ambivalence and a Preoccupied Stance

A different kind of adaptation occurs in response to a family life with inconsistently available parents and can yield a sense of anxiety about whether or not others are dependable. This response to inconsistent or intrusive parenting can yield a feeling of ambivalence and uncertainty. This may be experienced by an adult as a desperate need for others and a simultaneous sinking feeling that one's own needs can never be met. There may be a sense of urgency for connection that may ironically push others away and thus create a self-reinforcing feedback loop that others indeed are not dependable.

The pathway toward growth for aspects of adaptations that include such ambivalence and preoccupation often resides in a combination of self-soothing techniques, such as self-talk and relaxation exercises, along with open communication within intimate relationships. In some ways one can view this adaptation as involving an excessively active right mode with difficulty in the self-soothing that the right hemisphere specializes in. Memory and models of the self may not reassure the individual that his needs will be met and connections to others will be reliable. The sense of self-doubt at times may come along with a deep and nonconscious sense of shame that something is defective about the self. This sense of shame may be present in various forms in each of the insecure forms of attachment.

Understanding the mechanics of shame and how it may have been a part of our early life histories can help to free us from the ruts that these emotional reactions can create in our relationships with others. We may have developed layers of psychological defense that protect us from being consciously aware of what would otherwise be disabling anxiety, self-doubt, and painful emotions. Unfortunately, such defenses may prevent us from being aware of how these implicit emotional processes may directly influence our approach to our children. We may project onto them unwanted aspects of our own internal experience, such as anger at their helplessness and vulnerability. In this way, defenses that protected us during an earlier time may blind us from understanding our own internal pain and impair our ability to parent well.

Uncovering the layers of defense that may have been constructed in response to our own suboptimal parenting experiences is crucial in trying to make sense of our lives. Learning to calm the anxiety and doubt with relaxation exercises can be an important first step to learning other strategies to deal with our sense of discomfort. Having had intrusive and inconsistent parents may have blocked the development of our strategies for self-soothing. Learning "self-talk" techniques can be a very effective approach to taking care of yourself. Talking to yourself with clarifying statements such as "I feel this uncertainty now, but I am doing the best I can and things will work out" or "I am feeling nervous about what she said but I can ask her directly and find out what she meant" are examples of how the left-mode language can be useful to calm the right-mode anxiety. For those with layers of defense that may cover up an internal state of shame, it may be useful to recall that the belief that the self is defective is a child's conclusion, arising from noncontingent connections with parents. Realizing that "I am lovable" is important and can take the place of internal thoughts such as "I am not loved" or "I am unlovable." Finding ways that work to help the right hemisphere learn to self-soothe are the keys to growth for this form of adaptation. You can give yourself the tools that your parents were not able to offer to you as a child. In many ways, this is parenting yourself from the inside out.

### Disorganization and Unresolved Trauma or Loss

Adults whose experiences with their parents produced states of alarm and terror may have responded with internal disorganization. The sense of disconnection, from others and from one's own mind, may lead to a process of dissociation that might include a sense of being unreal or internally fragmented. More subtle aspects of disorganized adaptation may include becoming frozen under stress or feeling a rapid shift in one's state of mind in response to some interpersonal interaction. Unresolved trauma or loss can make fragmented internal experiences such as dissociation more likely to occur. When they do happen, such states may be more intense, frequent, and disorganizing and may make a repair of the disconnection with your child more difficult and less likely to occur with dependability. Finding a way to resolve these unresolved conditions is healing for both the parent and the child.

Unresolved issues may reflect the inability of the mind to integrate various aspects of memory, emotion, and bodily sensation into a spontaneous set of free-flowing and flexible responses. Such impairment to integration, sometimes involving a process called "dissociation," can be seen as states of rigidity or of chaos, which may be experienced by a parent as "stuck" repetitious patterns of behavior or as "flooding" states of overwhelming emotion. Freedom from these extremes of excessive rigidity and disabling chaos emerges as we move toward healing.

Though the events of our childhoods may have made little sense at the time, it is still possible now to make sense of how they have influenced us. As this integrating making-sense process brings together elements of the past with reflections on the present, you may find that your sense of possibility and your ability to construct a more flexible and enriching future is greatly enhanced. This process can be done in solitude, but it is often helpful to allow others to bear witness to our pain, and our journey toward healing.

Resolution depends upon the ability to be open and face what at times may seem like unbearable feelings. The good news is that healing is possible. Often the hardest step is acknowledging that there is some serious and frightening unresolved business. When we can take the deliberate steps to face the challenge of knowing the truth, we are ready to begin the path toward healing and growth and become more the parent we'd like to be.

**INSIDE-OUT EXERCISES**

1. Set aside some time to respond to the questions for parental self-reflection (pages 146–47). After waiting for at least a day, return to your written responses and read them aloud to yourself. What do you notice? How do your responses feel to you? How do you wish your parents might have offered you a different experience of being parented? How have these experiences shaped your own attitudes toward and interactions with your child? What are the most important lessons you have learned from this reflective process? Our life stories are not fixed and sealed in cement. They evolve as we grow and continue the lifelong process of making sense of our lives. Let yourself be open to your own lifelong development.

2. Without taking time to think about your answers in advance, write for several minutes completing the sentences "A good mother is . . ." or "A good father is . . ." Now, read what you have written aloud to yourself or to someone you trust. Is there anything on the list of responses that you feel you received from your own parent?

There may be many; there may be few or none. Of these responses, which qualities do you feel you give to your own children at this time? Choose one quality that you would like to develop initially as an area of growth.

3. Reread the discussion in "Adult Attachment," above, of the categories of attachment in children and adults and "Attachment Categories" (pages 136–38). Now write down your thoughts about how these classifications relate to your own experience of being a child. Let your mind feel free to select aspects from all the categories—you don't have to place yourself or others in one particular category. How do you feel that these patterns of communication have shaped your development? How have they influenced your approach to others? In what ways have they affected your intimate relationships in the past and how do they shape your ways of relating to your children and others in the present?

4. Are there particular leftover issues that emerge in the process of self-reflection? Do these themes influence your relationship with your child? Are there elements of your past that are particularly difficult to think about? Do you get a sense that there is a deep issue, such as fear of closeness, a shameful sense of being defective, anger at your child's helplessness, or another hidden core emotional process that is hard to talk about and may be affecting your relationship with your child? Do you feel that there are unresolved issues of loss or trauma in your life? What impact do you think these have on your internal experience and on your interpersonal connections to your child?

### The Adult Attachment Interview

Mary Main and her colleagues' important research instrument, the Adult Attachment Interview (AAI), was first created to attempt to understand why the parents of the children with differing infant attachment patterns acted in such distinct ways. To accomplish this, the researchers hypothesized that perhaps something in the parents' experience of being parented had a direct impact on their approach to their children. How could they determine the accuracy of this hypothesis? They decided to pose questions to the parents whose children's attachment status they knew. They took the most helpful topics and put them together into a formal interview that reviews the parent's recollections of his or her own childhood history. They found that these interviews allowed them to predict the interview subject's child's insecure or secure form of attachment to the subject with around 85 percent accuracy. This is very high for this type of study. In later projects, couples were interviewed and predictions were made regarding the yet-to-be-born child's attachment to each parent at one year of age. Again, correlations with the predictions made on the basis of the AAI data were high. The AAI clearly taps into some powerful aspect of parents' minds that directly influences their relationship with their children.

The AAI consists of about twenty questions that require that the subject search for autobiographical memories while being interviewed by a researcher. The researcher Erik Hesse has described this process as one that requires a dual focus: internal scanning and the external discussion at hand. The responses are taped, the tape is transcribed, and the written transcript is analyzed by a trained AAI coder who assesses the subject's reports of his or her childhood occurrences. But the principal focus of the analysis is the "coherence of the transcript"—a measure of *how* the individual tells the story of her early life history. This part of the analysis is an assessment of the coherence of the subject's discourse—her conversational interaction—with the researcher. Not having evidence for what is being said, saying too little, saying too much that is off the subject, or getting disoriented in the production of a response are all considered "violations" of discourse and can be quantified and coded. These violations are thought to indicate various forms of incoherence in the narrative of the subject. A careful quantitative analysis leads to an overall profile of the subject. The adult is then given a classification—"free, autonomous"; "dismissing"; "preoccupied"; and "unresolved/disorganized." As we've seen, the AAI classification is the most robust predictor of a child's type of attachment to that parent.

You may be wondering how such an assessment of a conversation, of the telling of an autobiographical narrative from one person to another in response to specific questions, could be an accurate assessment of the person's past. After all, memory is subjective. Why would a subjective accounting have such predictive power? Furthermore, narratives of the past are constructed from a blend of what happened, what we wished could have happened, and what we try to forget actually occurred. They are often biased by what we want others, and ourselves, to think about who we are. This is certainly true, but it's not all that relevant, because the AAI assessment is not dependent on the assumption that the subject's early family life happened in exactly the way described. More important is *how* the narrative is told, how it reveals the ways the individual has made sense of life as reflected in how he or she is able to integrate the various forms of memory into a coherent whole. A coherent narrative clearly reveals how an adult has come to make sense of his or her early life history.

One exciting aspect of the AAI research is the "earned-secure" attachment category in which described events appear to be very difficult, but the person has come to make sense of his or her life and can offer a coherent telling of the life story. Alan Sroufe has suggested that individuals who have come to make sense of their lives in the face of adversity have had a supportive relationship—such as with a relative, caregiver, teacher, or friend—that has served as a source of resilience. These findings reinforce the view that attachment is open to change and ongoing development.

*       *       *

### The Study of Earned Security: Making Sense in the Face of Adversity

The history of our current understanding of those with earned-secure states of mind with respect to attachment reveals an important aspect of the Adult Attachment Interview. The AAI was developed to understand some of the features that might be at play in determining why different individuals parented their children in the distinct manners to yield the various forms of secure and insecure attachments. Asking questions of individuals that required recollection and the narrative description of their personally lived events was not assumed to yield a necessarily historically accurate accounting of past experiences. The assessment of the interview is primarily based on the evaluation of the coherence of the story, not on the accuracy of its contents.

In the early 1990s, Mary Main and Erik Hesse would discuss in the formal training of the AAI researchers their thoughts about a group of individuals who had appeared to overcome adversity and achieved coherence in their narrative transcripts. The first formal study, published by Pearson and colleagues in 1994, compared those with such an "earned-secure" status to those individuals classified as "continuous-secure." That study defined earned-secures as those individuals with coherence of the AAI but narrative content descriptions that were "negative" in that they depicted insensitive, harsh, or in other ways significantly suboptimal parenting experiences that often would have been associated with various forms of incoherent AAI assessments and hence an insecure adult state of mind with respect to attachment. By the retrospective accounting of the AAI format, these adults appeared to have earned their security. The initial study found a statistically higher level of depressive symptoms in this group of people, but their children nonetheless were securely attached to them. Following up on this finding, Phelps and colleagues did a more extensive assessment of earned security by going into the homes of a study group and observing parent-child interactions in the various groups. That study, published in 1998, revealed that even in the face of stressful situations, such as bedtimes and getting off to school, those with earned security did extremely well. They could not only "talk the talk" by having a coherent AAI, but they could "walk the walk" of having sensitive, attuned caregiving even in the face of stress.

In 2002, an important addition to the research literature was published that deepened our understanding of the notion of earned security. Alan Sroufe and colleagues' important longitudinal study, spanning nearly a quarter of a century, has followed a large group of individuals from infancy onward. At nineteen years of age, members of the group were given AAI's, and an analysis of this data was used to examine earned security not just from the "retrospective" view (the only approach available at the time for prior studies given that the AAI was given to the parents of the young children in those investigations) but now also from the "prospective" approach (see Roisman et al, 2002). What they found was remarkable. Retrospective earned security was assessed using the same research criteria as the earlier studies (coherence with negative historical details) and was again found to be associated with increased levels of depressive symptoms. Now the researchers had the early observational data, and they found that as infants this group actually had sensitive maternal caregiving! However, their mothers had significantly higher degrees of depression themselves.

Next, they explored earned security from the prospective approach, defined as those individuals who had measured insecurity of attachment in infancy but who, as nineteen-year-olds, had a coherent AAI. These individuals had documented "prospective earned security." But the remarkable finding was that there was little overlap between this group of prospective earned-secures and the retrospective earned-secures! These findings need to be replicated in future studies, but it is an important initial observation reminding us of the importance of the AAI as a narrative assessment, not a historical document. Keeping in mind that AAI researchers do not make the assumption that the AAI is a historically accurate retelling, this finding reinforces the core notion that it is the coherence of the telling of the adults' story that is the most robust predictor of the child's security of attachment to that adult. Adult security—continuous or earned—as measured in narrative coherence is the feature most associated with a child's security of attachment.

The researchers identify a number of issues that may explain why retrospective and prospective earned-secures in this study were found to be in different groupings. The limitations of any study are crucial in interpreting the meaning of its findings, and Sroufe and colleagues note that their study involves data about maternal-child interactions, not assessments of fathers. Their study also does not have observational data from four to twelve years of age, a period of time that actually is a major focus of the questions of the AAI. It is possible, the researchers state, that retrospective earned-secures describe negative events because these in fact did occur but were not observed in the unavoidably limited observational focus of the study design.

Another possibility they explore is that the depressive in state of mind of these individuals naturally highlights the negative. This process, possibly reinforced or even initiated by either genetic or experiential factors related to the documented higher depression in mothers of these individuals, would explain a negative description in the AAI even though observed sensitivity of the mother during infancy was actually high. Another way of looking at this, the researchers propose, is that despite depression, these mothers were able to provide sensitive caregiving that offered a form of resilience that enabled their children to develop a secure attachment observed in infancy and later revealed in the coherent AAI at nineteen years of age.

Even though these young adults do not yet have children, we can get an idea about their interpersonal relationships by assessing the quality of their romantic relationships. In continuous earned-secures (documented security in infancy and a coherent AAI), prospective earned-secures (documented insecurity in infancy and a coherent AAI), and retrospective earned-secures (coherent AAI but negative historical description), the romantic relationship assessment revealed sensitive, attuned relatedness. This capacity for contingent communication documented in adult-to-adult relationships can be further assessed in these individuals in the future as they go on to have their own children.

Why don't the prospective earned-secures have negative historical details that would have them meet the same criteria as the initial earned-security group, now defined as retrospective earned security? Would alternative criteria yield different overlap of the retrospective and prospective groups? Would AAI's given at an older age yield more reflective comments than they do from the average nineteen-year-old who might be less willing to acknowledge his dependency on and vulnerability to his parents? Would a teen prone to depression be more likely to complain about parents? Clearly these possibilities would influence the outcome of this analysis.

If these findings are taken at face value as valid, then the researchers of this study suggest that perhaps we should not even use the term "earned" for the retrospective group, as they seem to have had secure attachments as infants. Nevertheless, they may have had difficult times later in childhood. One important finding that may be helpful in assessing the meaning of these findings is that the prospective earned-secures do not have higher degrees of depressive symptoms. Depression may produce a negative bias in recollection. This may explain the findings in the retrospective group. It may be that as the prospective earned-secures move from insecurity with their mothers toward security as young adults within their newly forming relationships in their lives, their narratives become coherent because they have worked through the suboptimal experiences, and their autobiographical recollections are now more "in the present" and not particularly focused, especially as a nineteen-year-old, on the impact of earlier events on their present life.

A finding from memory research may be helpful here. Autobiographical memory studies reveal two relevant processes: recency and reminiscence. Recency is how we tend to recall events that occurred in the recent past. Reminiscence is the process, which seems to begin around thirty years of age, in which we tend to recall our youth more than recent events. Could it be that the AAI, usually given to adults beyond their teens, is more influenced by reminiscence in the older parents being assessed? Will the AAI's of the longitudinal study be consistent between their teen years and later into their twenties and thirties, as these individuals become parents themselves? We will have to await such studies of this and other long-term investigations in the coming years.

A general issue about the continuity of the security of infancy (assessed with the "infant–strange situation" procedure) and that of adulthood (determined by the narrative coherence of the AAI) is relevant here in understanding change processes. When relationships remain constant, there is a strong correlation between infant and adult status. Security in childhood tends to be associated with security in adulthood; childhood insecurity is associated with adult insecurity. However, if relationship situations change, such as with loss, abuse, or change in sensitivity of caregiving, infant security can end up as adult insecurity. Also, insecure infants can end up as secure adults (these are the prospective earned-secures). While science cannot yet tell us exactly how a given individual will move toward security under normal nonclinical situations, the impression of numerous researchers is that it usually involves a sensitive, responsive, caring relationship with another person in that individual's life. Such positive connections may serve as a source of resilience and help people make it through difficult times.

A final point about earned security is important to mention. Mary Ainsworth carried out an assessment of the AAI that revealed that the reporting of a history of trauma or of loss by itself did not predict a negative outcome for the children of that parent. It was the finding of disorientation and disorganization in the AAI during discussion of trauma or loss, leading to the classification of unresolved, which was the most robust predictor of the disorganized form of attachment. Here again, the important point of the AAI is underscored that it is not what happened to an individual by itself that matters, it is how that person has come to process those events. Having some supportive connection with another person may help to make sense of difficult experiences in the home. Whether a person had insensitive caregiving, a depressive but sensitive caregiver, or experiences of trauma or loss, the important feature is how parents have come to make sense of their lives that matters most.

### Adult "States of Mind with Respect to Attachment"

Attachment research uses the term "state of mind with respect to attachment" to describe an adult's classification. Understanding some features of this aspect of our mind's functioning may be useful.

- Though we may have had several attachment figures in our lives, the AAI finding gives us only a single state of mind classification. The idea behind this is that during adolescence we coalesce our experiences into a single classification—most likely influenced by our attachment to our primary attachment figure.

- The idea of a state of mind is that there is a process by which our minds organize a stance, approach, or mental set that serves as a filter for our perceptions, biases our emotional responses, and directly influences our behaviors. Such an organizing process filled with a certain theme is characteristic of any general state of mind and its mental models; those specific to attachment can be quite tenacious and enduring.

- From a brain point of view, we can say that such a mental process is "ingrained" in patterns of neuronal firing in which past experiences and the adaptations that were generated in response to them create synaptic connections that are retained in memory. In the case of attachment, the model overlaps with a form of implicit memory: it is embedded early in our lives, is activated without a sense that something is being recalled, and directly influences our perceptions, emotions, behaviors, and bodily sensations. The mental models inherent in such learning are at the heart of what John Bowlby called "internal working models" of attachment.

Transforming attachment would involve changes in these states of mind. As with any change, learning may involve unlearning old patterns while simultaneously creating the conditions and experiences that promote the learning of new approaches. The brain can alter synaptic connections with the learning that accompanies new experiences. The brain may also be able to grow new neurons, especially in the integrative regions that may be at the heart of how a state of mind organizes such widely distributed processes. We can propose, then, that new self-understanding (itself based on new levels of neural integration) along with new interpersonal experiences that promote a new form of communication and connection can combine to enable one's state of mind with respect to attachment to develop toward security in adulthood. Openness to change may require a combination of new self-understanding and willingness to try new approaches to connecting with others.

## Emotion, Memory, and Attachment

The finding in the AAI that individuals in the classification of "dismissing" state of mind with respect to attachment reveal few details of their early family histories raises some intriguing issues about emotion, memory, and attachment relationship experiences. In the course of responding to questions in the AAI setting, these adults repeatedly affirm that they do not recall their childhood experiences. Subsequent research by van Ijzendoorn and colleagues attempted to discover possible cognitive problems, such as memory deficits or intellectual difficulties, that might be at the root of this lack of recall. These studies found no generalized memory or intelligence problems. The ability to recall other aspects of early life experience, such as factual recollection of television shows popular at the time, was intact. The intelligence of people in the "dismissing" group was normally distributed, as in the other groups.

These attachment researchers also looked for indicators of genetic causes of the AAI findings, including the variables that had been shown to have a genetic basis by research on identical twins raised apart. They found none. Variables that have some genetic component, such as intelligence, certain personality features, and lifestyle preferences and aversions were found to have zero correlation with the results of the AAI. This finding supports the hypothesis that attachment is primarily shaped by relationship experiences, not genetics.

Why do adults with the "dismissing" state of mind have such impaired recall of their family lives, not merely for the period of normal childhood amnesia but also, it seems, involving the recollection of details of lived experience during childhood? Although researchers cannot be certain that these individuals are merely not sharing what they actually do remember, clinically we see that this lack of recall appears to reflect an inability to retrieve memory. In general, impaired retrieval can be due to blocked access to stored memory, or to lack of encoding. We don't yet know definitively why this pattern is seen. However, we do know from research on memory, emotion, and the brain that an experiential process may explain this attachment research finding. In memory research, studies into emotions and recollection have shown that no emotional arousal may be associated with a low level of explicit memory encoding, storage, and later retrieval; excess emotional arousal may impair explicit encoding and thus block storage and later retrieval; optimal arousal facilitates memory processes such that later retrieval is made more likely.

What is optimal emotional arousal? From the point of view of effectively regulating emotions, this is a state in which the appraisal centers of the brain, many of which are located in the limbic circuits such as the amygdala and orbitofrontal regions, become activated to a degree that enhances neural functioning and neural plasticity. Optimal neural functioning means that these convergence areas maximize the integration of the relevant circuitry for processing the information at the time of the experience. Enhancement of neural plasticity means that the neuromodulatory chemicals released by these limbic appraisal circuits facilitate the creation of new synaptic connections. For example, the hippocampus may be a crucial "cognitive mapper" as it integrates the input from a wide range of neural areas into a cognitive whole. Optimal neuromodulation would facilitate such an integrative memory process.

Neuromodulatory circuits enhance the creation of new synaptic connections. By releasing neuro-active chemicals that facilitate the firing of neurons and also the activation of the genes that will lead to the production of proteins necessary for new synaptic formation, neuromodulatory circuits enhance learning. It has been hypothesized that experiences that are optimally emotional may involve the engagement of these neuromodulatory circuits in the encoding of memory that has a higher likelihood of being recalled in the future. The "storage strength" of these units of memory, these stored representations of experience, is stronger when the emotional circuits are optimally engaged.

One of the most important aspects of memory involves forgetting. If we were bombarded by recollections of most of our experiences, we would "lose our minds." Our minds require that we selectively forget most of what we experience, and the mind may have an automatic process whereby emotionally unengaging experiences are not encoded with much strength and are not easily recalled with much detail later on.

The proposal here regarding attachment experiences and the "dismissing" grouping is that the emotionally unengaging family lives of these individuals when they were children did not enable the laying down of easily recalled details of their childhood family experiences. There may be factual knowledge—of television shows, sports events, and facts about family events—but little autobiographical recollection. A unique feature of explicit autobiographical memory is that it has a sense of self and time. It appears to be mediated in a different set of circuits, those primarily within the right hemisphere, than explicit semantic/factual memory, within the left hemisphere. In this manner, the insistence on lack of recall, mediated by being raised in an emotional desert, creates an underdeveloped right-hemisphere autobiographical-knowing process, or autonoesis.

Endel Tulving and colleagues in Toronto have demonstrated that autonoetic consciousness—self-knowing awareness—is mediated by the prefrontal regions, specifically the right orbitofrontal cortex for autobiographical recollection. Tulving's view is that this right-side process enables the creation of a state of mind of self-knowledge and the experience of mental time travel that links past, present, and future. Extending these ideas to attachment research suggests that the experience of such self-understanding may be initially created by family experiences, but then may reinforce itself by the ways in which self-knowledge shapes the depth of subsequent interpersonal relationships. It may be that people with the capacity for compassionate understanding of the self are also equipped to focus that compassion on their own child.

Attachment research suggests that children engage their teachers in interpersonal interactions similar to those they have experienced with their parents. Attachment studies, as we've discussed, support the idea that attachment is a product of experience and is not a constitutional feature of the child. These two findings support the view that the patterns repeated outside the home reveal how an adaptation adopted by a child in response to a home environment is carried with her as she engages with the world at large. Later responses from others may then more deeply reinforce these adaptations and perpetuate the initial adaptive process and ingrain more strongly that pattern of being in the world.

If the development of the neural machinery necessary for self-understanding is limited, the capacity to have a rich inner life and to relate to the inner lives of others may also be quite limited. This limitation may be a useful adaptation to minimize the individual's vulnerability to emotional distress and disappointment. These are primarily right-hemisphere processes that may be functionally shut down. Movement toward security for these individuals would likely involve the activation or development of these underutilized brain mechanisms. In these individuals, experiences are to be encouraged that can access the various dimensions of right-hemisphere functioning, including nonverbal communication, awareness of the body, appreciation of emotional states in oneself and others, autobiographical recollection, and attuning to and aligning with others' mental states. Providing the supportive encouragement for the gradual activation of these emotional and interpersonal processes may help the individual to learn to tolerate vulnerability and intimacy with trusted others. Such experiential learning could then help transform the individual's state of mind with respect to attachment toward security.

## Digging Deeper

M. J. Bakermans-Kranenburg and M. H. van Ijzendoorn. "The First 10,000 Adult Attachment Interviews: Distributions of Adult Attachment Representations in Clinical and Non-Clinical Groups." *Attachment and Human Development* 11, no. 3 (2009): 223–263.

P. Fonagy, H. Steele, and M. Steele. "Maternal Representations of Attachment During Pregnancy Predict the Organization of Infant-Mother Attachment at One Year of Age." *Child Development* 62 (1991): 891–905.

R. Fraley, A. Vicary, C. Brumbaugh, and G. Roisman. "Patterns of Stability in Adult Attachment: An Empirical Test of Two Models of Continuity and Change." *Journal of Personality and Social Psychology* 101, no. 5 (2011): 974–992.

K. E. Grossman, K. Grossman, and E. Waters, eds. *Attachment from Infancy to Adulthood: The Major Longitudinal Studies.* New York: Guilford Press, 2005.

C. E. Hamilton. "Continuity and Discontinuity of Attachment from Infancy Through Adolescence." *Child Development* 7 (2000): 690–694.

E. Hesse. "The Adult Attachment Interview: Protocol, Method of Analysis, and Empirical Studies." In J. Cassidy and P. R. Shaver, eds., *Handbook of Attachment: Theory, Research, and Clinical Applications,* second edition, pp. 552–598. New York: Guilford Press, 2008.

E. Hesse, M. Main, K. Y. Abrams, and A. Rifkin. "Unresolved States Regarding Loss or Abuse Have 'Second Generation' Effects: Disorganization, Role Inversion, and Frightening Ideation in the Offspring of Traumatized, Non-Maltreating Parents." In M. Solomon and D. J. Siegel, eds., *Healing Trauma: Attachment, Mind, Body, and Brain,* pp. 57–106. New York: W. W. Norton, 2003.

M. Main, E. Hesse, and R. Goldwyn. "Studying Difference in Language Usage in Recounting Attachment History: An Introduction to the AAI." In H. Steele and M. Steele, eds., *Clinical Applications of the Adult Attachment Interview,* pp. 31–68. New York: Guilford Press, 2008.

M. Main, E. Hesse, and N. Kaplan. "Predictability of Attachment Behavior and Representational Processes at 1, 6, and 19 Years of Age: The Berkeley Longitudinal Study." In K. E. Grossman, K. Grossman, and E. Waters, eds., *Attachment from Infancy to Adulthood: The Major Longitudinal Studies,* pp. 245–304. New York: Guilford Press, 2005.

J. L. Pearson, D. A. Cohn, A. P. Cowan, and C. P. Cowan. "Earned- and Continuous-Security in Adult Attachment: Relation to Depressive Symptomatology and Parenting Style." *Development and Psychopathology* 6 (1994): 359–373.

J. L. Phelps, J. Belsky, and K. Crnic. "Earned Security, Daily Stress, and Parenting: A Comparison of Five Alternative Models." *Development and Psychopathology* 10 (1998): 21–38.

G. I. Roisman, E. Padron, L. A. Sroufe, and B. Egeland. "Earned-Secure Attachment Status in Retrospect and Prospect." *Child Development* 73, no. 4 (2002): 1204–1219.

R. Saunders, D. Jacobvitz, D. Zaccagnino, L. M. Beverung, and N. Hazen. "Pathways to Earned-Security: The Role of Alternative Support Figures." *Attachment and Human Development* 13, no. 4 (2011): 403–420.

D. J. Siegel. "An Interpersonal Neurobiology of Psychotherapy: The Developing Mind and the Resolution of Trauma." In M. Solomon and D. J. Siegel, eds., *Healing Trauma: Attachment, Mind, Body, and Brain,* pp. 1–56. New York: W. W. Norton, 2003.

———. *The Developing Mind, Second Edition: How Relationships and the Brain Interact to Shape Who We Are.* New York: Guilford Press, 2012. Chapter 3.

H. Steele and M. Steele, eds. *Clinical Applications of the Adult Attachment Interview.* New York: Guilford Press, 2008.

L. Strathearn, P. Fonagy, J. Amico, and P. R. Montague. "Adult Attachment Predicts Maternal Brain and Oxytocin Response to Infant Cues." *Neuropsychopharmacology* 34, no. 13 (2009): 2655–2666; see doi: 10.1038/npp.2009.103; published online 26 August 2009.

M. H. van Ijzendoorn and M. J. Bakermans-Kranenburg. "Attachment Representations in Mothers, Fathers, Adolescents, and Clinical Groups: A Meta-Analytic Search for Normative Data." *Journal of Consulting and Clinical Psychology* 64 (1996): 8–21.

E. Waters, S. Merrick, D. Treboux, J. Crowell, and L. Albersheim. "Attachment Security in Infancy and Early Adulthood: A Twenty-Year Longitudinal Study." *Child Development* 71 (2000): 684–689.

N. S. Weinfeld, L. A. Sroufe, and B. Egeland. "Attachment from Infancy to Adulthood in a High-Risk Sample: Continuity, Discontinuity, and Their Correlates." *Child Development* 71 (2000): 695–702.

# CHAPTER 7

# How We Keep It Together and How We Fall Apart: The High Road and the Low Road

## INTRODUCTION

Most parents love their children and want to give them a happy childhood, but they may become confounded by the complex dynamics of the parent-child relationship. "I don't want to yell and scare my children, they just push my buttons and I get so mad I can't seem to stop myself," a parent may say. Parents are surprised by their own behavior when they react more harshly toward their child than they ever imagined they would. Emotions may take over and get the best of us at times. When the parent-child relationship triggers a parent's unresolved issues, it is time to reflect on what internal process may be creating the external disconnection. Parents have an opportunity to heal and to stop being controlled by emotions from the past intruding on their present experience with their child.

When you are feeling stressed or find yourself in situations with your child that trigger past unresolved issues, your mind may shut off and become inflexible. This inflexibility can be an indication that you are entering a different state of mind that directly impairs your ability to think clearly and maintain an emotional connection to your child. We call this a low mode of processing. When in a lower mode of processing, which we'll call the low road, you may become flooded by feelings such as fear, sadness, or rage. These intense emotions can lead you to have knee-jerk reactions instead of thoughtful responses. When emotional reactions replace mindfulness, you're on the low road and it is very unlikely that you will be able to maintain nurturing communication and connection with your child.

In a low-road state of mind you are trapped in a repetitive cycle that ends up being unsatisfying to both you and your child. Having leftover and unresolved issues makes you vulnerable to entering the low road, especially under stressful conditions. If you have difficulty with separation, bedtime can become a nightly battleground. Although you may start out well, with a bedtime routine of reading stories, talking about your child's day, and giving goodnight hugs and kisses before you tuck your child into bed, the moment you leave the room your child calls you back, or gets up to come and get you. If you are ambivalent about leaving your child, you may have difficulties in setting boundaries. Your child's resistance to going to bed may precipitate a low-road response in you if your continued efforts to appease your child fail to work. If you become angry and yell or act aggressively, both you and your child may become distressed, and separation becomes even more difficult. After what can be hours of conflict, both you and your child end up unhappy, exhausted, and out of relationship. When a parent is angry with her child, the child can have difficulty in separating and falling asleep.

No parents really feel good about themselves or what they may have done to their children in this low-road state of mind. And yet when parents have not reflected on their own experiences they can return again and again to the low road. Part of the reason for this may be that when we are actually in that lower state, it is hard to bring ourselves back to a higher mode of processing. Reflecting on the origins of our issues can increase self-understanding and create a resilience that can minimize our entering the low road.

The high mode of processing uses a part of the brain called the prefrontal cortex, which is located at the front of the top part of the brain; hence we'll call it the high road. When processing on the high road, we are able to involve the rational, reflective thought processes of the mind. This supports our ability to reflect on possibilities and consider our choice of action and its consequences. The high road allows us to make flexible choices that support our values in raising our children. This doesn't mean that there won't be conflict or that our children will never be upset or unhappy. It does mean that we can choose how we respond to their behavior. Taking the high road gives us an opportunity to be thoughtful and intentional in our communication and choose actions that support a healthy, loving relationship with our children.

## THE LOW ROAD

It's the low road that challenges our parenting ability. A parent's unresolved issues can produce disorganization in her mind as well as her actions, which may cause her to respond with emotional intensity and unpredictability in her interactions with her child. She may even, without intending to do so, behave in ways that are frightening and confusing to her child.

Imagine that you are a three-and-a-half-year-old boy enjoying an outing at the park with your mother, who has been playing with you and delighting in your exploration of the playground equipment. You are feeling loved and valued. She tells you that it's time to go just as

TABLE 9. FORMS OF PROCESSING

HIGHER MODE, OR HIGH ROAD

A form of processing information that involves the higher, rational, reflective thought processes of the mind. High-road processing allows for mindfulness, flexibility in our responses, and an integrating sense of self-awareness. The high road involves the prefrontal cortex in its processes.

LOWER MODE, OR LOW ROAD

Low-mode processing involves the shutting down of the higher processes of the mind and leaves the individual in a state of intense emotions, impulsive reactions, rigid and repetitive responses, and lacking in self-reflection and the consideration of another's point of view. Involvement of the prefrontal cortex is shut off when one is on the low road.

you are climbing up the ladder to the slide. One of your mother's friends walks by and they begin a conversation. After several more trips down the slide your mother is still talking and you head for the jungle gym and manage to reach the top, where you proudly wave to your mother. She looks up at you, then at her watch, and becomes instantly furious because she is now late for an appointment. She screams for you to "come down this instant." She's shaking her finger at you and her face looks very angry and you wonder what happened to the mom that was enjoying being with you just a short time ago. Not wanting to connect with a "mad mom," you slide down the pole and hide in the crawl-through tunnel. Your mother reaches in and pulls you by your arm, yanking it hard and hurting you. Her voice and facial expressions have moved from anger to fury and she continues berating you for being such a "bad boy" and not listening. You start to

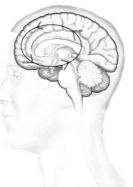

**Higher Mode or High-Road Processing: Integrated Functioning**

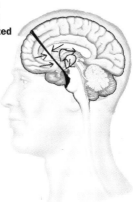

**Lower Mode or Low-Road Non-Integrated Processing**

FIGURE 4

cry and try to push her away from you. Her anger intensifies as she continues to yell at you for hitting her. She pulls you out of the tunnel and, ignoring your tears, hurries toward the car still yelling at you.

This mother's quick reaction was not a reaction to her child's behavior but came from her own issues. Perhaps noticing she was late for her appointment triggered a leftover issue of not being able to set boundaries for herself. Possibly when this mother was a child, her own mother didn't have the capacity to sufficiently meet her needs. Instead of being emotionally cared for, she would take care of her mother's needs and abandon her own. Now she was instantly triggered when her child wasn't taking care of her need to be on time for her appointment. When parents have unresolved issues, they may behave in ways that are frightening and confusing for their children. There are far more extreme examples of frightening experiences than this one, but even a parent's suddenly enraged face can be quite disorienting for a young child. A parent who is the source of alarm is placing the child in a conflictual experience and the child cannot make sense of the parent's behavior. A child is then faced with a stressful paradox that he cannot solve: the parent to whom the child needs to turn for comfort is now the source of fear. The child is emotionally stuck and confused and his behavior usually deteriorates.

The conditions that elicit such low-road reactions in a parent may resemble relationship situations or traumatic experiences from the parent's own past. A parent is especially vulnerable to being triggered and heading down the low road during many everyday moments in parenting when he or she is responding to a child's testing of limits, attending to a child's distress, or negotiating bedtime and other separations.

The low-road experience has four elements: a trigger, transition, immersion, and recovery. Triggers initiate the activation of our leftover or unresolved issues. Transition is the feeling of being on the edge, just before we fully enter the low road; it can be rapid or gradual. Immersion in the low road can be filled with intense emotions including the frustration and out-of-control feeling of being stuck in the low road itself. This lower-mode state shuts off the more flexible processing of the higher parts of the brain necessary for compassionate communication, so finding ways to recover from the low road is an im-

## TABLE 10. ELEMENTS OF THE LOW ROAD

TRIGGERS: Internal or external events that initiate the beginning of the low-road process.

TRANSITION: The movement from the integrated, higher mode of processing toward the depths of the low road.

IMMERSION: Being on the low road. The higher-road processes of self-reflection, attunement, and mindsight become suspended.

RECOVERY: The process of reactivating the integrative processes of the high road. High degree of vulnerability to reentering the low road may be present during recovery.

portant challenge to maintaining healthy relationships with our children.

Frequent and unacknowledged low-road behaviors on the part of the parent may directly produce fear and confusion in the child. The parent is often confused as well when she experiences internal conflict, ambivalence, or intrusive emotional memories that lead to rapid shifts in her states of mind. Sometimes a parent can become totally disconnected from interacting with her child as she becomes absorbed in her own internal processing in response to stress and she may have difficulty dealing with her own and her child's emotions. When the parent is in a low road of processing she can't respond to her child in an effective way. The best thing to do if a parent becomes aware of her intense anger and aggressive behavior is to stop interacting with her child. Until a parent has calmed down, the situation will probably get worse. The parent will most likely continue to become more out of control and the child will become more frightened.

## STUCK ON THE LOW ROAD

Dan was working with a family in his practice where the father had abrupt shifts into dramatically different states of mind that would often occur when he felt rejected. They were especially triggered when his daughter didn't comply with his requests. The father reported that he felt a "crazy feeling" as if "something was about to pop." He would sense a trembling in his arms and a pressure in his head and feel like he was going to explode. He said that he felt as if he was going out of his mind, as if he were entering a tunnel, receding from the world and drawing away from the people around him. At this moment he had entered the low road and could no longer stop the process. He knew that his enraged face was tightly drawn and the muscles of his body were stiff. Sometimes he would yell furiously at his daughter. At other times he would become filled with a rage he could not find a way to control and would squeeze her arm very hard or hit her.

The father felt ashamed of these outbursts and tried to deny his sudden and repeated shifts into this frightening and rage-filled state. His shame prevented him from engaging in any repair process with his daughter after such terrifying interactions. These repeated disconnections without repair produced within his daughter a view of an unreliable and confusing relationship with her father. Her memories of these frightening experiences may emerge as she grows older and be revealed as sudden shifts in her own state of mind, bursts of rage, or images of her enraged father. She may have the general sense that when she needs something, others might become irritated and betray her. Although this is not her father's intent, these are all a part of what she is learning from these experiences with him.

Why would this father act this way toward his daughter whom he dearly loves?

When the father was a child, he had been repeatedly subjected to the alcoholic outbursts of a drunken and angry father during which he was sometimes chased and beaten. His mother was depressed and withdrawn and unavailable to protect him, so he was extremely vulnerable to his father's unpredictable behavior.

As an adult, the father found it difficult to be receptive when his daughter would insist on things being done her way, as children often do. It was normal for his daughter to become irritated with him when she didn't get what she wanted. Children do that. But he felt that her challenging behavior was a personal rejection of him and this sense of rejection set off a cascade of abrupt shifts in his state of mind and he began to rage. This was his entry into the low road of processing.

How does this low road of processing happen? How does an unresolved issue make us vulnerable to entering the low road? Let's walk through the steps of the process to gain insight into these important mechanisms of disconnection. The perception of his daughter's irritation with him induces a shift in the father's mental state. This shift becomes linked with the representations connected with seeing an irritated face. These linkages activate a number of the father's unresolved issues. The emotional meaning of feeling rejected and the unintegrated implicit memories from past experiences flood his mind: behavioral impulses to flee, perceptual images of his enraged father or depressed mother, emotional responses of terror and shame, and bodily sensations of tension and pain. These linkages are made quickly and enter his consciousness without the feeling that he is recalling anything. As implicit memories, they are experienced in the here and now, as part of his present reality, and shape his internal experiences on the low road.

The father's perception of his daughter's behavior initiates an automatic cascade of implicit memories. This flooding rapidly shifts his state of mind. This sudden shift can lead to a discontinuous experience of the flow of consciousness, called dissociation. Individuals with unresolved trauma or loss might be especially vulnerable to such abrupt shifts and be more prone to entering the low road. At times, such a shift may appear as the entrance into a frozen, foggy state of mind. At other times, this shift may lead to the sudden onset of agitation and explosive rage.

The father experienced a sensation of "going out of his mind" and a feeling as if "he was about to explode." He was overwhelmed with intrusive implicit memories and suddenly shifted into a childhood mental state filled with that old and all-too-familiar sense of fear, rejection, anger, and despair. His sense of disconnection and impotence was experienced as shame. He interpreted his daughter's irritation as anger at him and he felt humiliated. Before he could pull himself out of this avalanche of feelings, he had entered the low road and become enraged. This lower-mode state shuts off the more flexible processing of the higher parts of the brain. In this altered, dissociated state, he behaved in a way—terrifying to his daughter—that he would never intentionally choose. He was, literally, out of control.

This father's repeated entry into these states of mind as a child had allowed them to become a part of his personality. This disorganization of his internal experience directly shaped his interactions with his daughter, who in turn experienced the disorganization of her own internal world. He was stuck on the low road.

## FINDING A WAY OUT

This father was at a loss to understand why he behaved in such a frightening fashion with his young daughter. Initially in therapy he had a very difficult time acknowledging that these interactions with his daughter actually occurred. After listening to a brief explanation of the high and low roads of processing, he began to have the ability to reflect on his own internal processes from a more objective, "distant" place. This distance gave him a sense of safety and he was able to transcend the feelings of guilt and shame that had previously kept him from reflecting on the issues that may be the root of his harsh and aggressive behavior toward his child. With this new framework he began his work in therapy toward healing the past.

What emerged from his work was the story about his drunken abusive father and his experiences with his daughter. Though the story became coherent with time, it was first revealed as layers of confusing and often frightening feelings and images. An understanding of how implicit memory can remain fully intact in the face of impaired explicit processing provided a meaningful framework to support his movement toward a coherent story. The only explanation for him up to that point in time was that he was prone to going crazy with fits of rage. With an understanding of the brain, memory, and the low-road states, he could see that, in fact, his brain was shutting off a crucial self-reflective function. With an understanding of how the prefrontal regions allow for flexible responses in the higher mode of processing, he could understand what "losing his mind" actually meant—being shut off from the part of his brain that was capable of making rational, thoughtful, and flexible choices, losing access to clarity of thinking, and being really stuck.

With this knowledge he could begin to make sense out of his experiences. His historical autobiographical information had to be linked to his present-day experiences in order to create a coherent story out of his internal and interpersonal experiences. His story revealed that terror was part of both the present and his traumatic past.

### RESOLUTION OF TRAUMA AND LOSS

Our childhood experiences may have involved trauma and loss in some form. Resolution of trauma and loss requires an understanding of the low road and its connection to patterns of experiences from the past. The passing of unresolved issues from generation to generation produces and perpetuates unnecessary emotional suffering. If our own issues remain unresolved, there is a strong possibility that the disorganization within our minds can create disorganization in our children's minds.

It is important to recognize that each of us may have leftover issues that create vulnerabilities that don't become apparent until we raise or work with children. Our entry into the low road can expose our leftover issues as well as our unresolved traumas and losses. Although most parents at times enter the low road, unresolved trauma or loss may make it more likely that these states will occur more frequently and will be especially intense.

When caring for children, it is inevitable that our own leftover issues will become activated in our minds. Even if we are not fully immersed in lower-mode processing, in which rational thinking is often suspended and intense emotions swell and bombard us like a tidal wave, leftover issues can still make it difficult to think clearly. They may bias our perceptions, alter our decision-making processes, and create obstacles to collaborative communication with our children.

Low-road interactions that are repeated and not repaired can impair the basic ABC's of attachment. Children need us to *attune* to them in order to achieve the physiological *balance* that enables them to create a *coherent* mind. Coherence is the state of mind in which the internal world is able to adapt to an ever-changing external world of experiences. Coherence enables an emerging sense of being connected to oneself and with others. Low-road experiences, often generated by leftover and unresolved issues, can disconnect parents from the very parts of their brains that enable attuned communication to occur. The children in that moment don't experience attunement and cannot move on to achieve balance or coherence at that point in time.

Unresolved issues often involve trauma or loss and can create more internal and interpersonal havoc than leftover issues. How do you approach such unresolved issues? If when thinking about loss or trauma you find yourself becoming disoriented or disorganized, you may want to reflect on those events and consider how they have impacted your life, your relationships, and the choices you have made. Start with the assumption that you, and your parents, have done the best that you can under the circumstances of your life. Instead of placing blame or judgment, be gentle with yourself. Be respectful of your bodily sensations, your emotions, and the images that come to your mind. Healing unresolved trauma and loss takes patience, time, and support. If you experience lack of resolution as a disorganization that at times involves intense emotions, confused thinking, impaired ability to express yourself to others, or social isolation, a qualified professional may be able to provide the support needed to help guide your process of healing.

No one goes through life without experiences of loss. Grief in response to the loss of a loved one is something each of us inevitably experiences. Healthy grief is a normal process. It allows us to have an internal reorganization of our relationship with our loved one after death. Continuing to care or think about a deceased loved one can be a part of healthy grief. However, our body and mind are not meant to actively grieve forever. Unresolved loss may reveal itself as continued intrusive thoughts and feelings about a loved one long ago deceased. Delayed or pathological grief can involve prolonged states of mourning that do not seem to reach a resolution. Grief can involve extended periods of social isolation and difficulty in ordinary daily functioning, owing to the intrusive, persistent, and overwhelming nature of unresolved loss. If continued self-reflection or participation in grief support groups does not help in gaining some sense of resolution, professional counseling may be helpful.

It is important to remember that children also respond with feelings of grief when they experience loss. Helping your child to process these feelings can help them make sense of these experiences so they do not continue to have a disruptive influence on your child's life. A child may experience grief not only with the death of a loved one but in many other situations that may not impact you in the same way. The loss of a caregiver, a change in availability of parents when there is a divorce, or even moving to a new house are very significant changes for a child and often involve a sense of loss. A parent can help a child make sense of his experiences by telling the story of the events from the perspective of the child's experience. When you retell the story, be sure to focus on your child's feelings, not your own feelings or the way you would like your child to feel.

When you reflect children's subjective experience, they are better able to process the loss. Reflective statements can be very helpful. Here are a few examples. If your young child has a change in her babysitter you might say, "Anna took care of you since you were a baby. You probably didn't want her to have to leave. Do you still wish you could see her every day?" Or after a divorce: "I bet it's hard when Mom and Dad live in different houses and you don't even get to decide where you want to sleep. What's hardest for you since Mom got divorced?" Or when the family moves to a new home: "It can be hard to get used to living in a new house. What do you miss most about our old house?" Children greatly benefit when we use concrete experiences, such as making books and drawings, to help them process their experiences.

A parent can help a child make sense out of confusing and fearful experiences of loss. What we might consider insignificant can be very important to a young child. The following story is an example of how a child's experience may be quite different from that of the parent.

A father went to the baby furniture store with his three-and-a-half-year-old son to pick up a mattress for a soon-to-arrive new baby. They were both enjoying sharing this experience and the little boy was feeling quite grown up. The father carried the mattress to the car just outside the entrance thinking that his son was right beside him. He put the mattress in the car and turned around to help his son into the car seat. His son briefly lost sight of his father because the mattress blocked his vision, and was standing with his back toward his father with tears rolling down his face. The child hadn't seen his father put the mattress into the car and thought he had been left at the store alone. The father reassured the child that he would never leave him alone and that in fact he had been right there the whole time. When they arrived home the little boy told his mother about how his father had "left him alone at the store." She carefully listened to the events and experiences of both of them and retold the story to her son, who was eager to make sense of this traumatic experience. After she had told the story several times the child seemed more settled and the issue seemed to be resolved. The mother ended with the statement, "If you ever have any questions about what happened, you can always ask me." Late in the afternoon as the mother and child were playing together, the child looked up and asked, "Did Daddy really leave me alone at the store?" He was still processing his feeling of abandonment, even though he had lost sight of his father for only two minutes and his father hadn't actually left him.

Children need time to process their feelings and make sense of their experiences. This child's emotional distress during those brief minutes created a state of emotional abandonment and terror that powerfully impacted his mind. By retelling the story, checking in with the child later about the fearful experience, and leaving the door open for future communication, parents can help their child make sense of an upsetting experience.


## PUTTING THE PIECES TOGETHER


When traumatic events are not resolved, they continue to impact our daily lives. With unresolved trauma, there may be a wide range of ways in which overwhelming experiences continue to influence you in the present. For example, when reflecting on feelings of having been threatened or frightened, you may sense a flood of emotions or a clouding of your thoughts. This may be a sign of lack of resolution. Lack of resolution can also manifest itself in the isolation of implicit from explicit memory. The various elements of implicit processing, including emotions, behavioral impulses, perceptions, and perhaps bodily sensations, can pervade your conscious awareness without your sensing that you are recalling something. These elements of implicit memory are reactivations of a past experience. It may feel as if you are having a fully formed, overwhelming experience. This is called a flashback. Or there may be an intrusion of only some fragmented elements of past experiences that do not form a sense of an entire event. Intrusions of disconnected bits of perceptions (such as sights without sounds), bodily sensations (pain in a limb), intense emotions (fear or anger), and behavioral impulses (such as to freeze or flee) may enter your awareness. They may give you no feeling that they are part of memory, of something from the past, but they may indeed be elements of "implicit-only" recollections.

In unresolved trauma, we may experience both "implicit-only" memories and fragments of explicit memory that have not been consolidated into a coherent life story. Explicit memories will give you the sensation of a remembered past event. If they are autobiographical, they will also have a sense of self and time. We may experience this nonintegrated explicit memory as bits and pieces of our past that do not yet fit into a larger narrative picture of our lives. Our life narratives emerge as our explicit processing integrates the vast array of elements of memory into the story of our lives. Reflecting on these elements of memory is crucial in order to resolve these disconnected fragments of our past traumas and weave them into a coherent life story.

Since reflection is often suspended when we are on the low road, we may first need to do the work of enhancing our recovery skills after an experience on the low road before we deepen our self-understanding. As time goes on, such reflection may become more possible at moments of transition into the low road, and perhaps even on the low road itself. Some people describe being able to observe themselves as if from a distance, even if they cannot change their behavior while on that low road. Being able to achieve this observational stance is an important beginning to freeing oneself from the prison that low-road immersion can become.

Our instinctual survival responses to fight, flee, freeze, or even collapse may become activated on the low road and dominate our behavior. The body's response may reveal these old instinctual reflexes in automatic patterns of response, such as tightened muscles in anger, an impulse to run away in fear, or a sense of being numb and immobilized. Becoming aware of our bodily sensations is a first step to understanding the experience on the low road. Making a conscious effort to alter our bodily reactions on the low road can help to free us from the prison of these ingrained reflexes. The brain looks to the body to know how it feels and to assess the meaning of things; thus, becoming aware of our bodily reactions can be a direct and effective means to deal with low-road immersion.

Changing the impact of the low road in our lives may require becoming familiar with the origins of these experiences and letting our minds move deeply into the layers of personal meaning that surround them. For example, for some the experience of being misunderstood or ignored may evoke a sudden feeling of shame, with its "pit in the stomach" sensation and a turning away from direct eye gaze. Understanding the trigger to that low-road, shame state would be important in liberating oneself from repeating such immersions. For others, being ignored may evoke an emotion of anger, which can create a transition into a state of low-road rage from which it may be difficult to recover. Understanding the specific triggers and how they evoke particular emotional responses is crucial in making sense of these experiences and then bringing resolution to such vulnerable issues in one's life.

Knowing about the brain can allow someone to move from self-judgment to self-acceptance. Our ability to enter into states of self-reflection often requires periods of solitude that can be quite difficult for parents of young children to achieve. Even a few brief moments reflecting at the end of the day or telling the stories of our experiences, especially those that have emotional content, to a friend can be beneficial. After an altercation when you were frustrated with your child's behavior and your own reaction to their behavior, you might want to ask yourself, "Why did I do what I did?" "Why did I think my behavior would lead to a positive change in my child?" These kinds of questions may help you develop greater self-reflection. Such mindful reflective experiences allow us to bring together elements of our internal worlds that may have been shut off for decades. A framework for understanding the brain and mind can deepen our self-reflective processes.

Bodily awareness and self-reflection may be followed by other experiences that enhance the healing process. Writing in a journal can be integrative and healing. Having trusted others bear witness to our pain and struggles can also bring a new sense of clarity and coherence to our lives.

How do you begin the process of healing? You can begin by writing or talking about your memories with a trusted adult who can support your growth in this process. Children can be burdened by a parent's experiences, and it is not their role to be your emotional support. If traumas were early, intense, and repeated, professional assistance is likely to be an important part of your journey toward healing and creating a coherent life story. Resolution of trauma and loss is both possible and important for you and for your children. Don't be paralyzed by your fear of confronting these unresolved issues. You don't have to be controlled by them anymore and they don't have to continue to impact the lives of your children.

**INSIDE-OUT EXERCISES**

1. Reflect on times when you have entered low-road states with your children. How have you acted at such times? How have your children responded to you when you were on the low road? Can you recognize the sensation when you are leaving the high road? Knowing your triggers and being able to recognize when you are entering the low road are the first steps toward changing the way they may be influencing your life and your ways of relating to your children.

2. Are there particular interactions with your children that frequently bring you into a low-road state? There may be recurring themes that can enable you to understand your experiences with the low road. What interactions with your children fill you with overwhelming emotion, such as fear, anger, sadness, or shame? For some, feeling abandoned or invisible brings them close to the transition to the low road. For others, feeling incompetent drives them "crazy." What theme acts as a trigger and brings you to the edge? Try to deepen your understanding of that theme in your life. What elements of that personal theme make getting back to the high road especially challenging?

3. If you have already entered the low road, self-reflection may be very difficult. If possible, remove yourself from interaction with your children. Move your body, stretch,

walk. Watch your breathing. As you begin to calm, observe your internal sensations and interpersonal interactions. You may find certain "self-talk" techniques useful in helping to decrease the intensity of your feelings and behaviors. "I need to calm down." "I am on the low road and these feelings and impulses are not dependable." "I need a break right now. Don't take this further . . . Take a time-out." Even if these strategies don't lead to immediate recovery, they can often help to reduce the destructive impact of low roads on your children, and your sense of self. Eventually, self-talk and observational reflection on the low road itself can help decrease its intensity and duration and facilitate a gradual but more reliable road to recovery. With reflection and insight you have the freedom to limit the negative impact of the low road by choosing a new and flexible set of responses.

4. Consider the possibility of changing the patterns of the past. When an issue is activated and you are about to transition to the low road, be aware that an alternative pathway is possible. Take a breath. Count to ten. Stop and get a glass of water. Remove yourself from the situation by taking a time-out or "emotional break." Now that you have created some distance, reflect on what was happening. Let yourself see the roots in the past that have led to present responses. You no longer need to follow these

age-worn paths. You have a choice not to repeat the old pattern. How might you respond differently next time?

## SPOTLIGHT ON SCIENCE

**The Brain in the Palm of Your Hand**

Understanding how the mind is able to function as an integrated system that achieves a balance within the body and brain and within interpersonal connections with others requires that we examine in depth how the mind emerges from neurophysiological processes and from social interactions. As we've seen, the perspective of interpersonal neurobiology enables us to view the mind as a process involving the flow of energy and information. This flow is determined by the connections among neurons within the brain, as well as by communication between people. In contingent communication, a sense of balance is achieved between the inner neural creation of a self and the outer responses of the social world. But how does the brain actually function to achieve some form of balance within itself, the body at large, and the social environment?

To begin to answer this question, we can look more deeply at how the brain's anatomy, its structure, is related to what it does, its function. The emerging findings from neuroscience reveal an exciting vision of the relationship between brain structure and function. Although the brain is quite complex, with its billions of neurons and trillions of spiderweb-like interconnections, we can actually get a handle on how mental processes emerge from this neural complexity by looking at the general architecture of the brain.

The following model has been useful to professionals and parents alike in getting a sense of a map of the brain and how it contributes to the processes of the mind. It is especially useful in showing how the mind creates the high road in integrative states and the low road in disintegrated, disassociated states of functioning.

If you take your thumb and bend it into your palm and fold your fingers over the top, you will have a surprisingly accurate general model of the brain. We can create a model, as a lot of neuroscientists do, by dividing the brain into three major areas (the "triune brain" model of Paul MacLean) and exploring some of the interrelationships among these anatomically separate but functionally interconnected regions: the brainstem, the limbic regions, and the cortex.

Hold up your fist so that your fingernails face you. The middle two fingernails are behind the eyes in this imaginary head. The ears come out the side, the top of the head is at the top of your bent fingers, the back of the head corresponds to the back of your fist, and the neck is represented by your wrist. The center of your wrist represents your spinal cord coming up from your back. The center of your palm symbolizes the brainstem, which emerges from the spinal cord. The brainstem, the lowest area of the brain, is the most ancient part in evolutionary terms and is sometimes called the primitive or reptilian brain. It is the part of the nervous system that takes in data from the outside world by means of sensations from the body itself and the perceptual system (except smell) and plays an important role in the regulation of states of wakefulness and sleep. It also is

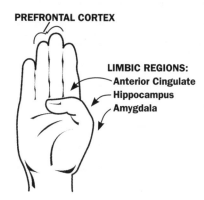

PREFRONTAL CORTEX

LIMBIC REGIONS:
Anterior Cingulate
Hippocampus
Amygdala

Place your thumb in the middle
of your palm as in this figure.

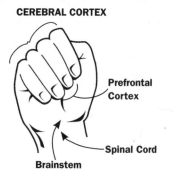

CEREBRAL CORTEX

Prefrontal
Cortex

Spinal Cord

Brainstem

Now fold your fingers over your
thumb as the cortex is folded over
the limbic areas of the brain.

FIGURE 5

important in mediating the major survival reflexes: fight, flight, freeze, or faint.

If you raise your fingers up and reveal the thumb curled into your palm, you will be looking at the area of your brain-model symbolizing the limbic structures, which mediate emotion and generate motivational states. These crucial limbic functions influence processes throughout the brain. Emotion is not limited to these limbic circuits, but appears to influence virtually all neural circuits and the mental processes that emerge from them. The limbic structures have similar evolutionary origins and similar neurotransmitters. Their influence is so extensive that the beginnings and endings of their structure as a system have been difficult for scientists to identify clearly. For these reasons, many modern neuroscientists are looking for another term

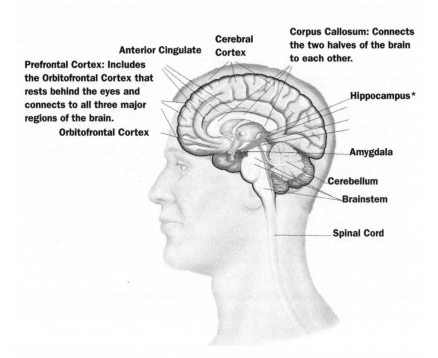

Prefrontal Cortex: Includes the Orbitofrontal Cortex that rests behind the eyes and connects to all three major regions of the brain.

Orbitofrontal Cortex

Anterior Cingulate

Cerebral Cortex

Corpus Callosum: Connects the two halves of the brain to each other.

Hippocampus*

Amygdala

Cerebellum

Brainstem

Spinal Cord

FIGURE 6. Diagram of the human brain looking from the middle to the right side. Some of the major areas of the brain are indicated, including the brainstem, the limbic areas (Amygdala, Hippocampus) and the Cerebral Cortex (with the prefrontal regions that for some researchers include the Orbitofrontal Cortex along with the Anterior Cingulate Cortex, both of which can also be considered as limbic structures).

*The shaded area represents where the hippocampus would be located on the other side of the brainstem in this diagram. At the head of the hippocampus is the emotion processing amygdala. Both of these structures are part of the medial temporal lobe.

besides "system" for these limbic regions. Much of the brain is divided into two hemispheres, and many structures such as the hippocampus are located in both hemispheres. For the sake of simplicity in this model, we will refer to the hippocampus rather than the right and left hippocampi, even though there might be a functional difference in the type of processing done on the two sides.

When a scientist says that a particular structure in the brain *mediates* a function, for example, the hippocampus mediates explicit memory, this means that various studies have suggested that for that particular function to occur (explicit memory), a healthy, intact structure (the hippocampus) plays an essential role. Playing a role often means that the neural processing of that region is either an essential building block (like visual processing of light and dark contrasts) or an overall process (like perception of an object). It may also mean that the region is carrying out an essential integrative function, pulling together the neural processing from other regions into a functional whole. As we'll see, the prefrontal and limbic structures often are essential in this integrative way.

There are several areas of the limbic region that are especially important to know for parenting: the hippocampus, the amygdala, the anterior cingulate, and the orbitofrontal cortex. Together, these structures enable the mind to carry out important functions that let a person create balance in the body, be adaptive to a changing set of environmental demands, and create meaningful connections to others. Our hypothesis is that through attachment relationships, the integrated connections of the child to the parent that honor differences and promote compassionate linkages, these structures achieve a level of integrative functioning within the brain that enables the child to thrive.

We've encountered the hippocampus in chapter 1, on memory. In your hand model, the hippocampus is the middle segment of your thumb. The hippocampus serves as a cognitive mapper, creating links among various widely distributed neural inputs that are important for integrating a number of processes that result in explicit factual and autobiographical memory.

The amygdala is the knuckle before the middle segment of your thumb, at the end of the hippocampus, and rests more deeply in the brain itself. The amygdala is important for the processing of a number of emotions, especially fear. "Processing" means both generating the internal emotional state and its external expression as well as making possible the perception of such states in the expressions of others. For example, the amygdala has face recognition cells that become active in response to emotionally expressive faces. The amygdala is one of the many important appraisal centers in the brain that evaluate the meaning of incoming stimuli. Research on this important area has revealed ways in which the amygdala influences a "fast track" of perceptual biasing, in which conscious awareness is bypassed to quickly alert the perceptual system to increase attention to threatening aspects of the environment. A "slow track" also exists in which emotional states such as fear are signaled to the consciousness-processing mechanisms of the higher neocortex. This setup enables us to sense danger and act on it rapidly without waiting for our more slowly processing conscious minds to be activated.

In our hand brain model, the anterior cingulate is in the middle two fingernails. In the brain itself, the anterior cingulate sits above the corpus callosum, the bands of tissue linking left and right hemispheres to each other. Some people think of this region as the chief operating officer of the brain. It helps to coordinate what we do with our thoughts and our bodies. It "allocates attentional resources," meaning that it determines what we pay attention to. The anterior cingulate also takes in input from the body itself, a process important in the creation of emotion. And the anterior cingulate plays an important role in our relationships, mediating a feeling of social rejection just as it does bodily pain. That's why our relationships deeply hurt us when they don't go so well in our lives!

The limbic circuits directly affect the functioning of an important structure in the brain not easily located in our hand model, the hypothalamus. Some authors include the hypothalamus as a part of the limbic circuitry. The hypothalamus is a crucial neuroendocrine center of the brain that initiates hormonal secretions and neurotransmitter flow involved in coordinating many brain-body functions, including the experience of hunger and satiety, and the response to stress. Also not easily represented in our model is the cerebellum, which is at the back of your hand where it connects with the wrist. The cerebellum is important for physical balance and also plays an important and newly discovered role in information processing. The cerebellum also sends inhibitory GABA (gamma-aminobutyric acid) fibers, which calm emotional irritability, to the hypothalamus and to the limbic structures. Of note is that childhood trauma and neglect have been shown to negatively affect the growth of these GABA calming fibers, and also to inhibit the growth of the corpus callosum and the brain as a whole.

The orbitofrontal cortex, the last of the limbic structures in our model, is also located in the third major area of the brain, the neocortex, which in our hand model is represented by your fingers closed over your thumb. Also known as the cerebral cortex or cortex, this region sits at the top of the brain and is generally regarded as the center of the most evolved brain functions: abstract thinking, reflection, and awareness that distinguish human beings from other animals. The cortex has a number of lobes that mediate distinct functions such as visual processing, auditory processing, and motor action. In parenting we are especially interested in the frontal part of the neocortex, called the frontal lobe. Represented by the front of your fingers from the second-to-last knuckles down to your fingernails, the frontal lobe mediates reasoning and associational processes. The front part of the frontal lobe is called the prefrontal cortex and is symbolized by the part of your fingers from your last knuckles down to your fingernails.

Two major areas of the prefrontal cortex are the side parts, called the dorsal lateral prefrontal cortex, and the middle parts, including the orbitofrontal cortex and the medial prefrontal regions as well as, for some, the anterior cingulate cortex and an area called the ventrolateral prefrontal cortex. The dorsal lateral prefrontal cortex, symbolized by the last segments of your outside two fingers, are the centers for working memory—the chalkboard of the mind—that enable you to remember a phone number long enough to dial it, or a sentence long enough to say it and remember what you said. The orbitofrontal cortex is called this because it's behind the orbits, or sockets, of the eyes. In the hand model, the middle two fingers from the last knuckles down to the fingernails stand for the orbitofrontal cortex and the other midline prefrontal regions. Some neuroscientists cluster the orbitofrontal cortex with its anatomically connected region, the anterior cingulate. Studies by J. Douglas Bremner and others have revealed, for example, that the anterior cingulate and the orbitofrontal cortex function as a circuit that may become quite impaired in their interactions with the hippocampus and the amygdala in the condition called post-traumatic stress disorder. As we'll see, the coordination of the prefrontal cortex with these other limbic structures is essential for flexible functioning. So for our purposes, it's helpful to simply think of the integrative role of these prefrontal and limbic structures working together as a team—on the high road, but falling "apart" and not working in a coordinated and balanced way on the low road. For practicality, we'll simply refer to the integrative role of the prefrontal region.

On your hand model, note how your middle two fingernails, the prefrontal cortex, connected of course to your neocortex (fingers), also rest on top of the other limbic structures (your thumb) and touch the brainstem itself (your palm). This anatomical position in your hand is the same for the actual brain! The prefrontal cortex is the only area of the brain that is one synapse away from all three major regions of the brain and even from the input from the interior of the body as well. It sends and receives neurons to and from the cortex, limbic structures, and brainstem, integrating these three areas into a functional whole. This unique position gives it a special role in integrating the complex system of the brain. The prefrontal cortex is the ultimate neural integrating convergence zone of the brain. If that were not enough, the prefrontal region also makes maps of others' minds, the input from other nervous systems. What this means is that five distinct sources of energy and information are coordinated and balanced by the prefrontal area: cortex, limbic area, brainstem, body, and the social world. That's what we mean when we say that the prefrontal region is integrative as it coordinates and balances these differentiated elements into a functional whole.

One of the many important functions that the prefrontal cortex is believed to carry out is the regulation of the autonomic nervous system (ANS), the branch of our nervous system that regulates bodily functions such as heart rate, respiration, and digestion. It has two branches, the sympathetic, which is like an accelerator, and the parasympathetic, which resembles a braking system. The two systems are regulated to keep the body balanced, ready to respond with heightened sympathetic arousal to a threat, for example, and able to calm itself down when the danger is past. The ability to have balanced self-regulation may depend on the orbitofrontal region's capacity to act as a kind of emotional clutch, balancing the accelerator and brakes of the body.

The prefrontal cortex also helps regulate the hypothalamus, the neuroendocrine center for the brain that sends hormones out into the body. Furthermore, states of alertness and emotional arousal, mediated by brainstem structures such as the reticular formation, are under direct influence of the cortex. Of note is that the orbitofrontal region is expanded on the right side of the brain. Many of these regulatory functions relate to the individual's stress-response mechanisms that also have been linked to a predominantly right-hemisphere function. We can see why the prefrontal region overall has been called the "chief executive officer"—because it helps to maintain balance in the body and mind through the integration of the three major areas of the brain and also coordinates them with metabolic states in the body as a whole and the input from our social world.

The prefrontal region is of particular relevance to parents because it integrates so many aspects of the brain that are central to good mental and emotional functioning. In addition to regulating the body through the autonomic nervous system, the prefrontal cortex is also involved in the regulation of emotion and emotionally attuned interpersonal communication, often involving eye contact. Along with the anterior cingulate and closely related regions, it appears to play a critical role in social cognition—the human capacity to sense other people's subjective experience and understand interpersonal interactions. Recent studies have also suggested that this region is crucial for the development of moral behavior. The prefrontal cortex also has to do with response flexibility, the ability to take in data, think about it, consider various options for responding, and then produce an appropriate response. Finally, it is believed that the prefrontal cortex and related regions are essential for the human creation of self-awareness and autobiographical memory.

The prefrontal cortex with its orbitofrontal region is the "highest" part of the limbic system, and the most "limbic" part of the neocortex. It is the end point for the input from the brainstem's autonomic nervous system, registering and controlling bodily functions. From this unique anatomical position, it achieves crucial neural integrative functions. At the level of the limbic structures, it is a part of an extended prefrontal circuit along with the anterior cingulate. This prefrontal region allows for attention to be organized along with the complex autobiographical, social, bodily, and emotional processes under orbitofrontal jurisdiction. As the orbitofrontal cortex interconnects with the hippocampus, it allows the processing of extensive cognitive maps of context and memory to create explicit autobiographical processing. As it links in with the central emotional processing of the fast-acting amygdala, the prefrontal cortex helps to shape our emotional states. Finally, the prefrontal cortex helps to coordinate our own states with those of others. In this manner, this integrative region is the gateway between interpersonal connection and internal balance.

At a minimum, staying on the high road requires the integrative prefrontal functions to be creating a flow of internal and interpersonal processes that enable flexible, stable, and adaptive functioning to be achieved. "Keeping it together" may literally depend on our brain's ability to integrate its functions. A well-tuned prefrontal region that interconnects with all the major areas of the brain may be an essential feature of doing well internally and interpersonally. The low road may occur when such integrative prefrontal processes are temporarily shut down. Such impaired functioning of the medial aspects of the prefrontal cortex with the hippocampus and amygdala have been shown to be present in clinical conditions of post-traumatic stress disorder. But even under less severe conditions, many of us may be prone to entering such states of impaired prefrontal integration given the right set of stressors and internal conditions. Such disconnection might occur with excessive emotional discharge from the amygdala or excess stress hormone secretion under the hypothalamus's neuroendocrine control. As these lower-mode states occur they may become ingrained in a form of implicit memory in which they become more likely to be activated. In other words, the neural firing patterns of a low-road, disconnected state may become more readily re-created when they have occurred with emotional intensity in the past.

In our hand model of the brain, you can envision the high-road state by enclosing your fingers around your thumb. This integrated system represents how the prefrontal regions can connect with limbic and brainstem mechanisms. The disconnected low-road state can be represented by raising your fingers; now, the prefrontal region can no longer carry out its integrative and regulatory functions with the other limbic structures. No longer coordinating with the anterior cingulate, attentional, social, and emotional mechanisms become dysregulated. The activity of the amygdala is no longer contained by the regulation of the prefrontal cortex and the emotions of fear, anger, or sadness may become excessive. Hippocampal creation of context may become lost as this region, too, becomes disconnected from prefrontal integration. Furthermore, the emotional clutch regulating the accelerator and brakes may become dysfunctional, and you are on the low road.

If the prefrontal region is impaired in some way or temporarily prevented from coordinating the activity of related regions such as the anterior cingulate, amygdala, and hippocampus, the individual may experience a sense of disconnection from others and a break-down in the reflective sense of self. The release of the normally inhib-iting prefrontal processes may result in the emergence of knee-jerk responses rather than flexibility of response. Trauma may impair the prefrontal region's capacity for neural integration, which is funda-mental to emotional recovery from traumatic events. The road to re-covery may likely involve new forms of learning that enable us to regain the prefrontal capacity to integrate bodily, emotional, self-reflective, and interpersonal experience into a coherent whole.

## The Low Road and the Prefrontal Regions

The low road of functioning is proposed as a state of processing in which the normally integrative functions of a higher mode, or "high road," are shut down. What evidence exists for the validity of this proposal? Teachers and therapists report that individuals raising chil-dren often describe times when they "lose their minds," "go nuts," "lose it," "fall apart," or "descend into the abyss." These subjective descriptions reveal a temporary change in a person's normally well-functioning mind. In this altered state, they no longer process infor-mation internally or behave externally as they usually would. Often, parents describe thinking and feeling in ways that become filled with rage, fear, or sadness that are "out of their control." Their behaviors may become rough, insensitive to their child's needs, and at times physically hurtful.

How do these changes in behavior and mental functioning occur? We can look to studies of the brain to make some educated proposals. The brain as a whole functions in complex ways based on how differentiated circuits become integrated into a functional whole. When neural integration is impaired, the normal functions of the brain that give rise to a coherent mind become disrupted. We can look to basic and clinical sciences that examine integrative functions of the brain for clues as to how the state of functioning might rapidly shift into a less integrated, less adaptive mode of processing. One of the key areas for neural integration is the prefrontal cortex.

Support for the existence of a low road of mental functioning and the role of the integrating functions of the prefrontal system is suggested by Mesulam (1998, 1035–1036): "Their pronounced anatomical connectivity with the amygdala suggests that the paralimbic sectors of the orbitofrontal area might play a particularly prominent role in the emotional modulation of experience. . . . In fact, damage to the orbitofrontal cortex can cause severe disruptions in the linkage of experience to the appropriate emotional state and can lead to extensive impairments in judgment, insight and comportment." These findings taken together with the role of the adjoining anterior cingulate suggest that functional disruption in these integrative areas of the prefrontal cortex would produce significant changes in the individual's internal experience and external behavior.

The interconnected functions of the orbitofrontal and the anterior cingulate cortices may be vital in creating a high-road state of flexible functioning. The orbitofrontal region of the prefrontal cortex has extensive connections to the cortex itself, to all of the limbic structures, and down into the brainstem. It regulates the branches of the autonomic nervous system (ANS), which balance bodily functions such as heart rate, respiration, and gastro-intestinal functioning. The anterior cingulate cortex has two components, one that regulates the flow of information (a "cognitive" area) and serves as a chief operating officer of attention, and a second area that takes in bodily input and creates emotional states and emotional expressions (an "emotional" area). Devinsky, Morrell, and Vogt have stated (1995, 279; 285; 298), "Overall, anterior cingulate cortex appears to play a crucial role in initiation, motivation, and goal-directed behaviors. . . . The anterior cingulate and its connections provide mechanisms by which affect and intellect can be joined. The cingulate gyrus may be viewed as both an amplifier and a filter, interconnecting the emotional and cognitive components of the mind. Complex social interactions such as maternal-infant interactions involve a level of brain organization beyond that subserved by simple sensorimotor reflex arcs. Several of these interactions may involve affect and executive functions subserved by anterior cingulate cortex and long-term memories stored in posterior cingulate cortex."

In cases of anterior cingulate dysfunction resulting from brain damage, a number of marked changes resembling more severe but qualitatively similar aspects of the proposed low-road state have been reported. Devinsky and colleagues note: "Behavioral changes following anterior cingulate cortex lesions included the following: increased aggressivity . . . emotional blunting, decreased motivation . . . impaired maternal-infant interactions, impatience, lowered threshold for fear or startle responses, and inappropriate intraspecies behaviour" (ibid., 285). Studies in monkeys and in hamsters reveal that anterior cingulate lesions disrupt the mother's ability to attend to the infant. The experiences of a brain-damaged patient described in the clinical literature revealed that "the social consequences of combined anterior cingulate and orbitofrontal cortex damage can be devastating. . . . [After a clinical challenge assessment] there was a disconnection between the intellectual understanding of the images and the autonomic expression with anterior cingulate cortex and orbitofrontal lesions. . . . The cingulate and orbitofrontal cortex are important in linking emotional stimuli and autonomic changes, and the behavioral changes that follow such stimuli" (ibid., 292). Brain damage to the medial portions of the prefrontal system can lead to marked changes in emotional and social functioning that are more persistent and severe but qualitatively similar to the transient changes proposed for a low-road state.

Seizures in the cingulate region are a neural process that provides clinical evidence to support the concept that the prefrontal regions can be temporarily disrupted in their functions. Cingulate epilepsy involves functional disruptions in coherent cingulate neural firing and reveals changes in behavior and subjective experience that are also consistent with our proposal for an altered state involving transient but significant changes in prefrontal functioning leading to a temporary low-road state. Devinsky and colleagues state, "Under normal conditions, behavioral responses such as vocalization in response to emotional stimuli are facilitated through circuits in anterior cingulate cortex and the adjacent motor area. Similarly, affective states are in part modulated by anterior cingulate cortex." When such anterior cingulate dysfunction is combined with difficulties with its middle prefrontal partners, the orbitofrontal cortex and perhaps also with the medial prefrontal regions, significant impairments in social understanding result: "Dramatic behavioral changes following anterior cingulate cortex lesions are associated with other lesions. Thus, when combined with orbitofrontal lesions, a devastating 'social agnosia' results" (ibid., 298). In other words, when these limbic-prefrontal regions are not able to function well, severe disruption to social interactions and understanding occurs.

Though we do not and may never have definitive studies of parenting experiences on the low road that are assessed, for example, in the setting of a functional brain imaging study, clinical findings in humans and research data in primates and other mammals reveal a set of changes in emotional, social, bodily, and regulatory functions that powerfully parallel what parents say happens to them under certain conditions with their children. Additionally, studies evaluating victims of trauma suggest that the medial aspect of the prefrontal region's functional connections with the important amygdala and hippocampus are also disrupted in post-traumatic states. Post-traumatic conditions, cingulate seizures, and brain damage are certainly all more extreme conditions than the experience of a parent entering the low road. However, stress may create a transient alteration in prefrontal ability to integrate a more coherent, high road of being that may have features of these more chronic and extreme clinical conditions. Furthermore, certain situations, such as unresolved trauma or loss, lack of social supports, or other psychosocial stressors may predispose a parent to more readily and deeply enter a low-road state.

A convergent approach to understanding our human experience draws upon these findings of science to suggest a view of what is likely to be going on inside our minds and brains. This view can help to deepen our understanding of what goes on inside us as we parent our children. Such a view helps us to move beyond the feeling that we are "losing our minds," beyond being paralyzed by guilt and shame because we feel that we have been imperfect parents. Instead, we can move toward a more compassionate understanding of ourselves and the importance of making a reconnection with our high-road states and with our children, who are waiting for us to return. As a mother in one of our workshops shared with us, knowing about the hand model of the brain and the idea of low-road behaviors, she came to realize that while these frequent moments of "flipping her

lid" weren't her *fault,* it was her *responsibility* to make a repair and try to reduce the occurrences of these nonintegrated states. Knowing about the brain can help you let go of self-blame and shame that make reconnecting with our children so difficult. Instead, we can realize that the brain just works this way, and we can use our minds to take responsibility for our behaviors and make a repair with our child.

## Digging Deeper

J. D. Bremner. *Does Stress Damage the Brain?* New York: W. W. Norton, 2002.

O. Devinsky, M. J. Morrell, and B. A. Vogt. "Contribution of Anterior Cingulate to Behaviour." *Brain* 118 (1995): 279–306.

M. Hamner, J. Lorberbaum, and M. George. "Potential Role of the Anterior Cingulate Cortex in PTSD: Review and Hypothesis." *Depression and Anxiety* 9 (1999): 1–14.

P. MacLean. *The Triune Brain in Evolution: Role in Paleocerebral Functions.* New York: Plenum Press, 1990.

———. "The Triune Brain in Conflict." *Psychotherapy and Psychosomatics* 28, nos. 1–4 (2010): 207–220.

N. Medford and H. D. Critchley. "Conjoint Activity of Anterior Insular and Anterior Cingulate Cortex: Awareness and Response." *Brain Structure and Function* 214, nos. 5–6 (2010): 535–549.

M. M. Mesulam. "From Sensation to Cognition." *Brain* 121 (1998): 1013–1052.

———. "The Evolving Landscape of Human Cortical Connectivity: Facts and Inferences." *Neuroimage* 62, no. 4 (2012): 2182–2189.

D. J. Siegel. "An Interpersonal Neurobiology of Psychotherapy: The Developing Mind and the Resolution of Trauma." In M. Solomon and D. J. Siegel, eds., *Healing Trauma*. New York: W. W. Norton, 2003.

———. *The Mindful Brain: Reflection and Attunement in the Cultivation of Well-Being*. New York: W. W. Norton, 2007.

———. *Mindsight: The New Science of Personal Transformation*. New York: Random House, 2010.

———. *The Developing Mind, Second Edition: How Relationships and the Brain Interact to Shape Who We Are*. New York: Guilford Press, 2012. Chapters 1 and 7.

———. *The Pocket Guide to Interpersonal Neurobiology: An Integrative Handbook of the Mind*. New York: W. W. Norton, 2012.

# CHAPTER 8

# How We Disconnect and Reconnect: Rupture and Repair

## INTRODUCTION

I t is inevitable that parents will experience misunderstandings, arguments, and other breakdowns in communication with their children. Such a breakdown is called a rupture. Parents and children often have different desires, objectives, and agendas that create tension in their relationship. Sometimes your children may want to stay up late at night playing games but you want them to get a good night's sleep. This will very likely create a limit-setting type of rupture. Other ruptures, such as when a parent is frightening to a child, are more toxic and create more distress in the child's mind. Although ruptures of various sorts may be unavoidable, being aware of them is essential before a parent can restore a collaborative, nurturing connection with the child. This reconnecting process can be called repair.

Parents need to be able to understand their own behavior and emotions and how they may have contributed to the rupture in order to initiate a repair process. Ruptures without repair lead to a deepening sense of disconnection between parent and child. Prolonged disconnection can create shame and humiliation that is toxic for the child's growing sense of self. For these reasons, it is imperative that parents take responsibility and make a timely reconnection with their children after a rupture.

Our minds are fundamentally linked to others' through the sending and receiving of signals. Ruptured connection, especially of our nonverbal signals, separates our primary emotions from the other person and we are cast adrift and can no longer sense our own minds within the mind of the other. We no longer feel felt, but instead feel misunderstood and alone. When this linkage with an important person in our lives is broken, our minds will quite likely experience a disruption in balanced and coherent functioning. We are not meant to live in isolation, but are dependent on one another for emotional well-being.

Sometimes relationships with children become filled with tension. Parents don't always like their children, or feel positively toward them, especially when their children are acting in ways that make the parents' life more difficult. Being compassionate toward your own emotional experience enables you to accept these challenging altercations with your children with less distress and self-recrimination. Sometimes, a parent's sense of guilt at her own anger toward her child can prevent her from being aware of, or even caring about, a ruptured connection. Unfortunately, this guilt can block the initiation of repair and deepen the distance between parent and child. Having self-understanding about these processes can open the important door to reconnection.

It can be difficult for a parent to provide structure and boundaries for his child while simultaneously offering collaborative communication and emotional alignment and connection. How can a parent achieve this? Balancing structure and connection is a basic goal we can move toward but is impossible ever to fully achieve. As parents learn to balance their own emotions without swinging between feelings of guilt and feelings of anger toward their child they are better able to give their child both nurture and structure. Being kind and empathic toward yourself can help you not get overly involved in your own emotional reactions to your child.

Understanding alone cannot prevent these disrupted connections from occurring. Some will inevitably happen. The challenge we all share is to embrace our humanity with humor and patience so that we can in turn relate to our children with openness and kindness. To continually chastise ourselves for our "errors" keeps us involved in our own emotional issues and out of relationship with our children. It is important to take responsibility for our actions, but not to condemn ourselves because we are not able to act in some idealized manner or because we are not further along in our own developmental process. We, just like our children, are doing the best we can at that point in time and like them we are learning more respectful ways to communicate. No matter how well we apply all the best principles of parenting, misunderstandings and disruptions in our connections to our children will inevitably occur. Disconnections are a normal part of any relationship. It is more helpful to use our energy to explore the possible routes to reconnection and see these times as learning opportunities rather than to belittle ourselves for what we think are our failings. Take a deep breath and relax! We are all learning throughout our lives.

## OSCILLATING DISCONNECTION AND BENIGN RUPTURES

The connection between child and parent is always changing. Sometimes the communication is contingent and collaborative and both

**TABLE 11. TYPES OF DISCONNECTION AND RUPTURE**

Oscillating disconnection

Benign rupture

Limit-setting rupture

Toxic rupture

parent and child feel understood. This alignment and joining feels good. When there are repeated experiences of connection, there can be the sense of resonance in which we feel the positive presence of another within us and sense that we are within the other.

But this ideal level of connection by its very nature cannot be sustained consistently. It is inevitable that there will be breaks in this wonderful sense of joining. These disruptions take many forms. In the day-to-day experience of living, both parents and children have oscillating needs for connection and solitude. Life is full of this tension between connection and autonomy. Attuned parents sense these cycling needs of the child and give space so a natural separation can take place and then make themselves available when a child needs to be close to them. Sometimes, a child's need for connection may feel intrusive to parents who want to have some time for themselves. However, a parent may first need to give time to a young child before she is able to find time for her own needs. Older children are better able to understand and tolerate the parent's need to be alone because they also experience more clear boundaries between their own needs for connection and solitude. Adolescents are a whole other story, often seeking isolation from parents and greater connection to their peers.

If you are feeling a need to be alone, it is more helpful if you can express that need directly to your child. Being able to say, "I need some alone time right now, but in ten minutes I will be able to read you that story" is better than trying to ignore a child or resent the child for "forcing you" to spend time with her. By letting your child know that your feelings and actions are about meeting your own needs and not a result of her behavior, seeking solitude will not automatically be experienced as a personal rejection by your child. Without this clarity about your own needs, you may try to create distance in less helpful ways, by getting angry at your child or thinking of your child as being "too needy."

Other forms of disruption include misunderstandings in which a parent does not "get" the messages being sent by her child. Perhaps the parent wasn't paying attention to what the child was communicating because the parent was preoccupied. Perhaps there was a lack of comprehension of what the signals meant. Children often don't say in words exactly what they have on their minds. Even if the message is ambiguous, the child still desires to be understood. A parent may focus on only the external aspect of behavior of the child, and miss the deeper level of meaning. Parents may themselves give incongruent messages, which greatly confuse children. Parents, too, don't always say what they mean and children try to sort out the true message beneath the conflicting signals.

All of these situations of benign rupture can be frequent in our daily lives with our children. When children are feeling emotional, whether they are excited or upset, they have a heightened need to be understood. It is at these times that even benign ruptures may be especially painful for children. Making repairs in a timely and caring manner is important if children are to build and maintain a sense of resilience and vitality.

## LIMIT-SETTING RUPTURES

Children benefit when parents create structure in their lives. A child learns which behaviors are appropriate within the family and the larger culture by the limits set by parents. Setting limits can create tension between parent and child. When a child desires to do something and the parent cannot allow that behavior, a limit-setting rupture may occur. Such a rupture at the time of limit setting involves the child's emotional distress and a sense of disconnection from the parent. In this situation, the child's desire to carry out a particular action or to have some object is not supported by the parent. This lack of attunement between parent and child can leave the child feeling distressed. The child wants something that the parent cannot give him. Parents can't always say yes to their children's requests. If children ask for ice cream right before dinner, demand a toy every time they go to the store, or attempt to climb on the dining-room table, there is a parental need to set limits. These limit-setting experiences are crucial for the child. They involve the child's developing a healthy sense of inhibition in which the child learns that what he or she wanted to do is not safe or socially appropriate within the family setting.

When a child hears "no," she feels a sense that her desire or action was "wrong." The parent can help to redirect her impulse into a more socially appropriate and safe direction. The key to staying in connection during these limit-setting interactions is to realign yourself with your child's primary emotional state. You can empathize and reflect back to your child the essence of her desire without actually fulfilling her wish: "I know you'd like to have some ice cream. It's too close to dinner, but maybe can have some ice cream after dinner." This is a much different experience for the child than just hearing the parent say: "No! You can't have it."

Many times, empathic and reflective comments can help your child move on past his frustration at not getting what he wanted. However, even if the parent offers the most supportive response, a child may still feel upset and adamant about his desire, no matter what you say or do. Allowing your child to have his distress without trying to punish him or indulge him can offer him the opportunity to learn how to tolerate his own emotional discomfort. You do not have to fix the situation by giving in or by trying to get rid of his uncomfortable feelings. Letting your child have his emotion and letting him know that you understand that it's hard not to get what he wants is the kindest and most helpful thing you can do for your child at that moment.

Parents can often learn how to parent more effectively from reflecting on unsatisfying or difficult experiences with their children. Here's one mother's story that might help you gain more understanding of the dynamics between a mother and child during a limit-setting rupture.

It's 7:30 a.m. and Mom is in the kitchen fixing breakfast while mentally reviewing the many tasks on her "to do" list for the day. Four-year-old Jack is his usual energetic self and starts to climb on some baskets that are stacked in the corner by the refrigerator. "Don't climb on those. It's not safe. What do you want up there anyway?" asks Mom.

"I want my bunny grass," replies Jack.

Mom definitely doesn't want to deal with the leftover Easter basket grass, so she lies and tells him that there is no bunny grass on top of the refrigerator. Jack, knowing she's not telling the truth, confronts her with "Yes there is!" Mom, feeling guilty about having lied, produces the bunny grass and reluctantly hands it to her son, asking, "What are you going to do with it?" Jack starts to pull all the grass out of the bag and heads toward the dining room. "Don't take that stuff out of the kitchen. I don't want it all over the house. It gets caught in the vacuum cleaner." Jack ignores her until she calls his name sternly and he turns back toward the kitchen. "Just pretending," he says as he goes over to his play kitchen and starts "decorating" it by laying bits of bunny grass on it.

Dad is reading the newspaper at the table. A few minutes later, Mom looks over and sees that Jack is now "decorating" the breakfast table. Each place mat, as well as the salt and pepper shakers, now has bits of green plastic grass draped across it. Mom views this as a big mess that *she's* going to have to take care of and says sternly, "Don't put any bunny grass at my place." Jack ignores her and "decorates" her place. "Most kids don't even get to have bunny grass when it's not Easter," she says, but Jack continues to ignore his mother. "You're not listening to my words," she chastises.

Dad tries to offer some support: "Your mother doesn't want bunny grass there." But Jack seems to have lost his hearing and continues with his play. In exasperation, Mom yells, "Get that bunny grass out of there!" Dad calls Jack's name in a threatening tone.

Jack, feeling angry at being yelled at, mumbles, "Oh, all right" as he clears the bunny grass off his mother's place and throws it on the floor. This blatant act of disrespect angers Dad, who jumps up and tries to wrestle the remaining grass away from his son. "That's it! No more bunny grass!" he shouts. Jack screams and cries and tries to hang on to the bag of grass, wailing, "But I did what you said! I took it off!"

The morning has deteriorated into a yelling match as Mom and Dad try to wrestle the bunny grass away. They have a strong sense of how foolish this all is. Jack is angry and getting more furious by the moment. Exasperated, the parents offer some ineffectual consequence as a "compromise" in which the bunny grass ends up in a "time-out" and is put in the cupboard. Later in the day, while his parents are away,

Jack talks his babysitter into letting him scatter the grass all over the house. "Sure, Mom lets me do it," he tells her.

How might the morning have gone differently? One obvious solution would be to pack the leftover bunny grass away if it was not an option to play with it. But twenty-twenty hindsight is always easy. There are many other points in the communication where the scenario could have moved in a more positive direction. Here are some possibilities. Mom could have told the truth about the bunny grass being on top of the refrigerator and set the limit right then. "Yes, the bunny grass is up there, but it's not for playing with now. Do you want to make a plan to play with it later when we're finished with breakfast?"

What if Mom had already given him the grass to appease him and soothe her guilt for lying, before she saw the impending difficulties? She could have stopped cooking breakfast and addressed the situation while it was still only a small irritation. "Jack, this isn't going to work right now! I should have said, 'No bunny grass until after breakfast.' I'm going to put it away now and you can think of a place where you can play with it later so it won't make a mess." By setting a limit early she can be more effective and follow through without frightening him or challenging him to engage in a power struggle.

We could imagine the scene with different options at points where the situation moves toward greater conflict. At these points, what could the parents have said or done differently? There is no single right answer, but many possible choices a parent could make. It is, however, important for the parent to take some action rather than just verbally reacting and threatening the child. As we can see, Jack kept pushing the boundary to see what was "enough," because the limits were set with ambivalence and there was a lack of clarity and congruence in his mother's messages. She was giving mixed signals that encouraged him to find out what she really meant so he persisted in testing the limits.

Reworking this scene, thinking up other choices and anticipating the possible outcomes, is an interesting exercise. You might want to think about a situation with your own child when you lost your temper and you weren't satisfied with the outcome. Try to understand why your child may have responded the way he or she did and what you could have done to shift the energy in a more positive direction.

We have to check in with ourselves so we can become clear in our own minds on the limit we want to establish and the message we want to deliver. Setting limits is a way of showing respect for ourselves as well as our children; they are more effective when we set them before we are angry.

## TOXIC RUPTURES

Ruptures that involve intense emotional distress and a despairing disconnection between a parent and a child can be experienced as harmful to a child's sense of self and are therefore called "toxic ruptures." Children may feel rejected and hopelessly alone during these moments of friction. When a parent loses control of his or her emotions and engages in screaming, name-calling, or threatening behavior toward a child, it can create a toxic rupture. Toxic ruptures often happen when a parent has entered the low road. For a person in a low-road state, flexible and contingent communication is impossible. These toxic ruptures are the most distressing form of disconnection for children because they are often accompanied by an intense feeling of overwhelming shame. When shame occurs, there is a physiological reaction; children may have an ache in their stomach, heaviness in their chest, and an impulse to avoid eye contact. They may feel deflated and withdrawn and begin to think of themselves as being "bad" and defective.

When parents have leftover or unresolved issues, they are especially at risk for entering into toxic ruptures with their children. Parents can become lost in the depths of a low-road state, and even if they recognize the toxic rupture, they will be unlikely to be able to repair it until they have centered themselves. This centering often requires that parents disconnect themselves from the interaction with their child. They needn't necessarily create physical distance, but making the mental space to become centered is usually vital in order for parents to calm themselves down. If parents stay on the low road and continue to try to interact, they will be emotionally reactive and their leftover issues will cloud their ability to parent effectively.

Sustained and frequent toxic ruptures may lead to significant negative effects on the child's growing sense of self. It is important that these ruptures be repaired in an empathic, effective, and timely manner so that the child's developing identity is not damaged.

Once we feel calm and are able to reflect on the situation, we will have left the low-road state. It may be difficult for any of us to see ourselves as having been hurtful or frightening to our children—but we can be. We may be reluctant to see ourselves as being out of control. This reluctance may lead to our denial of our own role in the toxically ruptured connection with our children. It is important for us to take responsibility for our actions: an important aspect of repair is to acknowledge our own role in the disrupted connection: "I am sorry that I yelled at you when you came in late for dinner without even listening to what you had to say. It was starting to get dark. I must have been worried that something might have happened to you. I didn't mean to scare you by yelling so loudly. I really went overboard. I should have listened to you and then told you what I was worried about." Reflecting with a child on the inner emotional experience of the altercation can be crucial for both parent and child. This reflection is often effective at repairing both the rupture and the child's sense of shame and humiliation at having been the recipient of our out-of-control, low-road behaviors.

Reflective dialogues, discussions between parent and child about the internal level of experience, focus on the elements of mind that contributed to and were created by the rupture. In this manner, the parent can reflect on his or her own internal experiences and reactions as well as those of the child. In the end, the goal is to achieve a new level of alignment where both child and parent feel understood and connected and have regained their sense of dignity so they feel good about themselves and the other.

Although the occurrence of toxic ruptures should be avoided, when they do occur we can use them as an opportunity for increased personal insight and enhanced interpersonal understanding. In the repair process the child learns that although the going might get rough at times, reconnection can be achieved, bringing with it a new sense of closeness with a parent.

## THE EXPERIENCE OF SHAME

An overwhelming feeling of shame is often at the heart of toxic ruptures for both parent and child. Feeling impotent to effect a positive outcome for our child's behavior can lead to a rush of feelings within us, including frustration, humiliation, and rage. A sense of being defective may accompany the feeling of shame and be related to our own history as a child. The ways in which we were misunderstood and mistreated in our childhood can return as ingrained states of mind that have mental models of ruptured relationships as well as automatic reactions that defend us from our own internally generated feelings of shame. Caught in the avalanche of feelings and defenses, we may easily lose sight of our children's needs as we enter the low road. When we are in that state, collaborative, contingent communication is often the furthest thing from our minds.

When in a shame state, as a parent we may become excessively concerned about the opinions of others around us and be filled with ideas of right and wrong. When children misbehave in public, we may become more focused on the reactions of strangers than on trying to understand the meaning of the behavior for our child and on effectively guiding her. We may treat children more harshly in that setting because we feel humiliated and are not attuned to our children's emotional needs. Such preoccupations may make it difficult to attend to the signals from our children. We may also become especially vulnerable to feeling that we are judged as being incompetent and ineffectual. If we feel we cannot control our children in the ways we think others expect we should, we may begin to feel ashamed of our sense of being powerless. For some, this feeling of shame may evoke leftover issues that activate an old pattern of rigid responses. We enter the low road as a cascade of defenses erupts against that feeling of shame.

A defense mechanism is an automatic mental reaction that attempts to keep us in balance by blocking our awareness of a disorganizing emotion. In this case, early adaptations to a feeling of shame may have included a defense that kept our conscious minds unaware of these painful early emotional experiences; this set of defensive mechanisms is called a shame dynamic. When it is activated we can become swallowed up in a torrent of old patterns of response that serve to keep shame hidden from our conscious minds. These reactions are both historic in nature and are current attempts to ward off our present awareness of the dreaded feeling of shame.

Shame dynamics, as with all defense mechanisms, occur without our awareness or intention. Our complex minds utilize these automatic processes in order to minimize disruptive thoughts and emotions that would interfere with our daily functioning. Bringing these dynamics to our awareness gives us the opportunity to increase the possibilities for how we live our lives and know our selves.

Shame also may play a central role for the child during toxic ruptures. At moments of intense emotion, the feeling of being disconnected may automatically, biologically, induce a state of shame. This may simply be a natural response to disconnection at a time when connection and contingency are desperately needed. If the ruptured state is prolonged, the shame may become toxic, meaning that it can be harmful to the child's sense of self. If the disconnection from the caregiver is associated with parental anger, the child may come to feel both shame and humiliation. These feelings lead a child to turn away from others, to have an intense sense of distress, and to have the cognitive belief that there is something defective about the self. Rigid defenses may develop, and these can directly shape the development of the child's personality. Repeated, prolonged, and unrepaired experiences of toxic ruptures are damaging to the child's developing mind.

If we have had repeated experiences of toxic rupture without repair in our own childhoods, shame may play a significant role in our mental lives, even outside our awareness. Sudden shifts in our feelings and in our communication with others may indicate that the shame defenses are becoming activated. Experiences in which we feel vulnerable or powerless can trigger the defenses that our minds have constructed to protect us from awareness of such a painful state of shame when we were children. These defenses may persist into our adulthoods and influence our parenting. Reflection is an important process in beginning to unravel the complex and rapid mechanisms of a shame dynamic that may be impairing our ability to parent effectively.

The signs that a rupture has occurred in communication with our children may be subtle or extreme. In its more extreme form rupture can lead to withdrawal or aggressive reactions on the part of our child.

Subtle forms of rupture may be revealed when the child looks away and avoids eye contact. There may be a shift in the tone of voice and a diminished sense of engagement, reflecting a possible state of shame within the child and in ourselves. In other situations, ruptured connection may cause the child and/or the parent to become focused on only a particular aspect of a discussion and not remain open to the full communication from the other. Not feeling heard, both parent and child may become more assertive in their points of view and an argument begins which leads to more disconnection. Leftover issues may contain themes of disconnection that are especially vulnerable to being reactivated during this cascade of intense emotions and then further impair communication as the parent rapidly enters the low road. This creates a feedback loop in which both parent and child feel ignored and less and less heard and understood by the other.

## REPAIR

Repair is an interactive experience that usually begins with the parent's own centering process. It's extremely difficult for the parent to make a repair if he or she is still angry and on the low road. The parent needs to take time to reenter the high road before beginning the interactive repair process. Only from that mindful, centered place can the parent begin the important and necessary process of reconnection. It is very difficult for the child to initiate rejoining with a parent who has become enraged or in other ways frighteningly disconnected to him. Recall that disconnection, especially at times of heightened emotions and the need for connection, produces shame. Because of this, repair can be difficult for both parent and child and for this reason may often be left undone. Some parents may want to "just get over" their moods and the unpleasant interactions and move on to act as if the rupture had never occurred; this leaves the child feeling an even greater disconnection from his own feelings.

Even if your own experiences as a child included toxic ruptures that your parents were unable to repair, you can still change what might be your natural impulse to "just forget" the rupture. Such denial of toxic disconnections may have given you leftover issues around the experience of shame. Now you have the opportunity to heal your own emotional issues and provide a different set of emotionally enriching experiences for your child. Instead of being experienced as dreaded burdens, the moments and issues of parenting that are the most challenging can be turned into opportunities for growth and renewal. Intimacy and resilience can be built through these rupture and repair experiences.

How does a parent become centered in order to begin the repair process? First, it is important to gain perspective on the interaction that caused the conflict by creating some distance mentally, and perhaps physically as well. All ruptures can't be solved immediately. Individuals may need different amounts of time to process both the events and their feelings about them. Taking time away from the interaction is important. Breathing deeply and relaxing the mind by not continuing to think about the rupture can help to create a more peaceful and calm state of mind. It may be helpful to do something physical in order to shift the intensity of the mood and utilize the adrenaline-driven energy in a nonharmful way. Moving your body can help to change your mood and give you a fresh perspective on the situation. Going outdoors and being in nature can have a calming influence on your mind. Getting a drink of water, making a cup of tea, or changing your physical location can often help you break out of a low-road rut.

Once you have achieved a calmer state of mind and have regained some sense of clarity, think about how to reconnect with your child. Do not do this prematurely because until the thunderclouds of your mind have lifted you are vulnerable to being out of control and can quickly slip back onto the low road. While on the low road you may say or do things to your children that you may regret later and would never have done in a higher-mode state. If possible, do not touch your child when you are in the low-mode state. Your fury may be expressed through your hands and turn into harmful actions more quickly than your mind can inhibit these low-road impulses. If your anger has been expressed by physically hurting your child the repair process becomes even more complicated and difficult, yet even more necessary to do in a timely manner.

Once you are back on the high road again, consider how you will approach your child. Think of your own leftover issues and why they got activated during this interaction. Be sure to let your awareness of these issues take on a dual focus: you need to understand your own emotional baggage and you also need to attune to your child's experience and signals. This dual focus is crucial so that you don't rapidly reenter the low road if your child initially resists your attempts to reconnect, which might happen. It is important to respect his or her sense of timing as well as your own.

After you have calmed and centered yourself, consider your own historical issues. How did the interaction activate a particular theme of your life history? How did your child's response trigger your low-road reaction? Try to see the interaction from your child's point of view. What do you envision to be his or her experience of this interaction and the rupture? It is easy to forget how small and vulnerable our children are, making their experience of the rupture much more intense and frightening. Children, especially when they are very young, cannot tolerate long periods of disconnection without feeling abandoned and distressed. It is important that we make efforts to reconnect with them as soon as is possible.

The anger parents may feel at themselves for acting out of control may keep them from making efforts at repair. A parent's own defenses to the child's emotional reactions may block the parent from seeing the child's need for reconnection. Some parents may hate their own vulnerability or emotional needs for connection and project that hatred onto their child via intense rage at their child's behavior. In this way a parent's leftover issues may block the repair process.

## INITIATING REPAIR

Being able to focus on both your own experience and that of your child is a central feature of effective repair. With that dual focus in mind, you can now begin the process of interactive repair. Getting on the same physical level of your child can be helpful in making a reconnection. Younger children often want close proximity to you; older children may feel intruded upon and prefer that you begin by keeping a little distance. Though children may not invite you to reconnect or may not bring up issues of the rupture, it is important for you as the parent to make a nonintrusive, empathic attempt to begin the repair process. Children with different temperaments have different ways of being on the low road or in a meltdown state. Some take a long time to recover; others recover quickly. Children often cannot recover until the parent has reached out to make a connection with them.

Learn and respect your child's style for processing a rupture and making a reconnection. Timing is important. If you feel rebuffed after your first attempt, don't give up. Your child wants to be back in a warm and positive relationship with you. It is the parent's role to initiate repair and you should find another time to initiate a reconnection. It is helpful to address the experience of the rupture in a neutral way, keeping your dual focus in mind: "This has been so difficult for both of us to be fighting like this. I really want us to feel good about each other again. Let's talk about it." You and your child have had a mutually shared experience of disconnection, even though each of you has had your own different perceptions of the events. Reconciliation does not happen if you are trying to place blame. As the parent, you have the responsibility to own your behavior and know your internal issues.

After beginning with your stated intentions to make a reconnection and your acknowledgment of the difficulties that you have been having with each other, you can listen to your child's feelings and thoughts. Do not interrogate him. Curb any tendency to judge the responses you hear. Just listen. Be open to his point of view. You do not need to defend yourself. Listen to your child's experience before you share your own experience of the interaction. Be sure to join with your child by reflecting back how you hear his experience of the events. Pay attention to both the content of the perceptions as well as the meaning of the emotional experience.

As you begin to discuss the toxic aspect of the rupture—the yelling, the cursing, the throwing of the ketchup bottle—it is important to state that sometimes people, even parents, have meltdowns in which they do not act rationally. People do temporarily "lose their minds" and then recover them. Children need to hear this so that they can come to understand people, their parents, the mind, and the nature of toxic ruptures. Without this understanding, they cannot make a coherent story out of these frightening ruptures.

Children of different ages and of different temperaments may tolerate these ruptures and efforts at repair quite differently. Infants are especially vulnerable to these toxic ruptures and don't have the skills to process what has occurred. Pre-school-age children may respond to parental low roads with confusion and acting-out behavior and may need more consoling and nonverbal connection than older children do. These younger children need more help in incorporating toxic ruptures into a coherent story through role-playing, puppets, storytelling, and drawing. More verbal older children may respond to discussions of what has occurred and be open to exploring their experience of the parent's behavior as well as their own responses.

## THE ACCELERATOR AND THE BRAKES

How we communicate with our children helps to shape the ways they learn to regulate their own emotions and impulses. Earlier in the book we talked about the important prefrontal region of the brain that helps coordinate a number of important processes such as self-awareness, attention, and emotional communication. The prefrontal region is also crucial in the process of regulating emotions. A part of this area is situated in a location such that it directly connects the three major areas of the brain and coordinates their functions: (1) higher thought processes of the neocortex, such as reasoning and complex conceptual thinking; (2) the motivational and emotion-generating limbic region in the middle of the brain; and (3) the lower, brainstem structures that bring in input from the body and are involved in basic processes such as instincts and the regulation of sleep-wake cycles and states of alertness and arousal.

The prefrontal region sits at the top of the part of the nervous system that regulates bodily organs such as the heart, the lungs, and the intestines. Many researchers believe that the signals from these parts of the body enter the brain and help determine how we feel. It turns out that the prefrontal region not only receives signals from these bodily systems but also serves as the "chief executive officer" and regulates their functioning. The prefrontal region has a "clutch" mechanism that helps balance the accelerator and brake functions. The prefrontal region controls the autonomic nervous system's sympathetic—accelerating—and parasympathetic—decelerating—branches. When the accelerator is activated, the heart speeds up, the lungs breathe more rapidly, and the gut starts to churn. When the brakes are applied, the opposite set of reactions occurs, and our bodies become calmer. Keeping the accelerator and brakes in balance is the key to healthy emotional regulation.

When a person is excited about something, the accelerator is activated. When we say no, the brakes get applied. You can try an experiment at home that illustrates this. Ask some friends or relatives to sit down and close their eyes. Ask them to just sit quietly and observe their internal sensations. Now say the word no and repeat it clearly and slowly five times. Wait a few moments and let them notice their reactions. Now say the word yes clearly and slowly five times. After giving them some time to reflect, ask for their responses. People often experience the "no" as a feeling of heaviness, withdrawal, and a vague discomfort. With "yes" there is commonly an uplifting, exhilarating, or peaceful sensation.

"Yes" is thought to activate the accelerator; "no" activates the brakes. In parenting we are frequently faced with the need to set limits. "No" becomes a response that children often hear from us, especially after their first birthday. An eighteen-month-old is excited about exploring her environment and now has the motor abilities to turn her desires into action. She will inevitably find something to explore that is dangerous and the parents will want to set limits on her exploration. When we set a limit the child's brain accelerator is activated and then the brakes are applied. In an ideal situation, as the brakes put a stop on her behavior, her accelerator will be released and she will listen to what we are asking her to do.

Purely in terms of brain functioning, an activated accelerator followed by the application of brakes leads to a nervous system response with a turning away of eye gaze, a feeling of heaviness in the chest, and a sinking feeling. This is similar to the profile for shame. This limit-setting "no"-induced form of shame is what some researchers call a "healthy" kind of shame, different from toxic shame and humiliation. Children learn to regulate their behaviors by developing an emotional clutch, located in the prefrontal cortex, that can turn the accelerator off when the brakes are applied and redirect their interests in more acceptable directions. Children learn that at times what they want to do is not permissible and they need to redirect their energies.

Children who do not have these important limit-setting experiences may have an underdevelopment of the emotional clutch, which is a building block of response flexibility. Parents who don't want to be identified as the "bad parent" often resist setting limits and are unable to provide their children with these important developmental experiences. Their child's emotional clutch is not developed enough to allow her to rechannel her energy in productive ways. One of our roles as parents is to facilitate our children's development of their ability to balance the brakes and accelerators that enable them to delay gratification and modify their impulses. This means that our children learn to hear "no" but also to maintain their spirit and belief in themselves. These are essential components of emotional intelligence. Your child is excited about throwing toys or climbing on the kitchen counters. We say "no." The brakes are slammed on. We help to redirect him toward actions that will satisfy both his energy and his desire for movement so we say, "You could go outside and get that basket of balls. I bet you could throw those really far," or "The counters aren't for climbing but you could go outside and climb to the top of the tower by your swing set. You could probably see really far from up there." Now the accelerator is turned on again when the child senses your attunement to his inner state of excitement about throwing or climbing; the brakes turn off as the accelerator now redirects play to socially appropriate activities. Setting limits, creating clear boundaries for acceptable behaviors, and offering structure gives children the important experiences that enable them to have a sense of safety and security. Experiencing these essential "no's" gives them the opportunity to develop the capacity for self-regulation that permits them to put on the brakes and redirect their energies in other directions. The emotional clutch of children who have not been given the chance to develop this important aspect of self-regulation often reveals an inability to flexibly adapt to the environment. "No" is followed by a torrent of indignation and tantruming as their prefrontal region cannot engage the clutch and create a flexible response. The ensuing meltdown and inflexible behavior is exhausting for child and parent alike.

Parents help teach children to regulate this emotional clutch in order to balance their accelerator and brakes. In order to do this, a parent needs to be able to tolerate the tension and discomfort that a child may experience when the parent sets a limit. If a parent cannot tolerate a child's being upset it is very difficult for the child to learn to regulate her emotions. A limit-setting "no" is best followed by calm, clear follow-through by the parent. If we always capitulate and give our child what she wants just to keep her from being upset, we will not support our child in developing a healthy ability to apply the brakes and redirect an activity. It is not necessary or even helpful to verbally reason with our children all the time. If we only value the logical mind, we'll get into endless arguments and negotiations, and our children will think that if they make a reasonable argument, we should always act according to their wishes. Sometimes it's okay to say, "No, that's just not okay with me" or "I understand how you feel, but I'm not going to change my mind." We don't have to explain all of our decisions or give a reason for everything we do and expect our children to readily agree with us.

If we scream and yell at a child when she complains after we say "no," we will generate the unfortunate response of deepening a sense of toxic shame and humiliation. With toxic shame, the child feels discon-nected from us, misunderstood, and as though his or her impulses are "bad" rather than misguided and in need of rechanneling. If a child also experiences anger from the parent, then the prefrontal region may have brakes applied (after "no") with continued accelerator application (in response to the parental anger). This is a toxic situation, like trying to drive a car with both the accelerator and brakes applied. The result of the child's emotional clutch failing to disengage the accelerator when the brakes are applied is a state of "infantile rage." The circuits are on overload and the child quickly enters the low road. At times, such an overload and low-road situation can happen in the parent as well.

## PARENTAL MELTDOWN AT THE TOY STORE

Dan tells a story about himself and his son that illustrates rupture and repair.

I had made a plan with my twelve-year-old adolescent son to go to the toy store to get an extra piece of hardware he had been wanting for his video-game set. The only time we could go that day was right be-fore a very important meeting. It left us with only half an hour for the trip to the store. I felt pressured and uneasy about the timing, but didn't want to disappoint him and said yes when I should have post-poned the trip for another day. We skipped lunch and went to the store and found the item we wanted. While the clerk was getting it, which cost twenty dollars, my son had a few minutes to look around at some newly released software and saw an expensive new baseball video game that he wanted to buy. I wasn't prepared to spend more time or more money and told him, "We've got to go and that's too much money." He wanted to use some of the sixty-five dollars he had saved from his allowance and doing extra chores, but I tried to convince him to consider a less expensive game instead. We argued about the merits of his choice of game, the value of money, and his not needing to have everything his friends had.

I was feeling hungry, pressured, and preoccupied about the meeting and irritated that he couldn't be satisfied with the games he already had and was asking for one more thing. I started lamenting to myself how modern American life is so filled with material goods that it's hard to help kids develop good values. Then I started to give him the parental lecture: "Well, forty dollars is a lot of money. You need to plan ahead of time for this kind of purchase. You need to appreciate the things you have; you can't buy everything you want. You can think about it this week and if you still want it next weekend, I'll bring you back here and you can buy it with your own money."

"I have money at home. And I have thought about it. I want to get it. I've earned enough money and you can't stop me from getting it," he said.

Responding to his threat I met his challenge sternly: "You are not getting this! Now let's go!"

He came back with "Fine. When we get home I'll tell Mom and she'll just bring me right back here and buy this for me."

"No, she won't."

"Yes, she will," he baited me. "You'll see, she makes the final decisions, not you. She'll bring me right back here and I'll have it."

Incredulous, I said, "No, she won't. You are not going to your mother to get this thing."

"Yes, I am," he taunted. "She'll just get in her car and bring me right back here."

"Now just stop saying that or you won't even get this piece of hardware that we came to get."

"I am going to tell her how mean you are. She'll bring me right back here now."

"If you say that one more time, we're just going to go home without this hardware."

"Fine. Mom will buy that for me too."

I slammed the hardware down on the counter. Now, entering my own total parental meltdown, I said, "That's it. We're out of here." I stormed ahead to the car. When we were driving home, he said through his tears that he guessed that I'm just a very sensitive guy and that sometime, somehow, when I'm not expecting it, he would get back at me.

This threat put me over the edge and I exploded on the low road. I began to swear at him and told him that he would not be allowed to play video games for the next ten months.

When we got to the house, he ran to his mother and told her that I had cursed at him and had been mean to him and he began to plead with her to take him back to the store.

He went to his room and I went to mine. My blood was boiling and I knew that I couldn't see straight and I could hardly contain my fury. I really needed to cool down and get centered. I took some deep breaths and moved around to try to get the tension out of my body. I thought of how I should either extend his consequences into the next year or just remove his video-game machine altogether. When I started to calm down I thought about my son and our ruptured connection. We had had a great morning playing together. I was excited to get him this piece of hardware. And then I thought about his face when he told me about the new baseball game. He was so excited. He was telling me about all the features of the game and how much fun it would be and that he would teach me how to play it so we could play together. I remembered being preoccupied about the upcoming meeting and feeling glad that I could just keep the plans we had made together and make the time to go with him to get the hardware item. I wasn't prepared to think about an additional forty-dollar purchase, even if it was his money. I gave him a series of mixed messages. If it was okay to buy a less expensive game, why couldn't he just use his own money and buy a newer and more expensive one? It really didn't make any logical sense, and he knew it, and even expressed that to me at the time.

But I wasn't open to listening to him. When he began to threaten to get his mother to countermand me, I sank fully into the low road. When I totally disregarded his need for autonomy (after all, it was his money), he fought back with his threats for maternal rescue. What a triangle. Not seeing the emotional meaning of all of this, I had stayed focused on the external elements: a "spoiled child" ungrateful for the hardware he was getting was now so disrespectful that he couldn't even accept a reasonable limit on further purchases.

Setting limits is important and enables children to learn frustration tolerance, response flexibility, and a balanced functioning of their emotional clutch. On the other hand, kids do need to be allowed to make their own decisions and learn from the experience of making some mistakes. Though it was totally reasonable to limit what my son bought, my inability to see his frustration, as expressed by his threats to "tell his mother" on me, led to my own meltdown. I could no longer parent effectively. I became a bag of reacting neurons merely lashing out with ineffective parental consequences and inappropriate language. I was totally immersed in the low road.

One way of viewing this situation was that my own accelerator and brakes were being applied simultaneously and I could no longer "drive the car" of my emotions. My not tuning in to his excitement about the game may have triggered his own accelerator-brakes overload condition and he began to plot his revenge. My response to this then paralleled his and neither of us could effectively communicate with the other.

After I reflected on the whole experience I wanted to reach out to reconnect. I went to his room and sat on the floor by the bed, where he was sitting in tears. I told him that I was sorry that we were fighting and that I wanted to make up with him. He just looked away, but stopped crying. I told him that what I said to him was wrong and that I wanted us to figure out what happened. He talked about the game and the fact that he had been thinking about it for a long time. Only I didn't know this, he said, and I should have let him purchase it. I told him my understanding of what happened, and about how I was preoccupied and what issues were flooding my mind at the time. I told him I now realized that he was so excited about the game and because of my own concerns I had missed sharing that excitement. He began to cry. I hugged him and told him what a buffoon I had been. I apologized for cursing at him. I told him that I had been in my own meltdown and that the consequence of ten months off all video games was extreme.

I said that now I understood and respected how important it was for him to decide how to spend his own money. I also expressed to him what I thought his contribution to the fight was, pointing out that he pushed the limit way too far by threatening to go to his mother and to "get back" at me. I said that I understood he was mad, but that he had just pushed things way too far even if he was mad. The same, I said, was true for me.

I said that his mother and I would talk later and come up with a plan on how to handle this whole thing. She and I later decided that we would remove the extreme consequences and put all purchases on hold for a week. Later that day, we had a family meeting to go over what happened. He and I told the story of the meltdown at the toy store and after some tears during the retelling, we had some pretty hearty laughs as we each role-played the other's behavior. He does a pretty good imitation of me.

## INSIDE-OUT EXERCISES

1.  How did your family experience ruptures during your childhood? Do you recall any benign or limit-setting ruptures? How did your parents handle them? Now, consider a time as a child when you experienced a toxic rupture. What happened? How did you feel? How did your parents behave? What was the outcome of that rupture? Was there a repair process? How did the process change your relationship?

2.  Reflect on a rupture with your own child. What happened? How did you feel? How did your child respond? Was a leftover or unresolved issue of yours activated? Can you recognize any patterns in ways this issue interferes with your ability to have a collaborative connection with your child? In this rupture was there an element of its being toxic? If you entered the low road, where did that path take you internally and in your actions? How did you recover from the low road? If you didn't engage in repair, how would you do that now and in the future?

3.  What aspect of repair is most challenging for you? What helps you identify and then move out of the low road? Can you sense when disconnection is occurring? Once it has happened, how are you able to remove yourself from the toxic interactions in order to become centered and reenter the high road? How do you approach the reconnection process? What aspects of communication in

the repair experience are most difficult for you? How does the feeling of shame play a role in the low road and in the possible impediments to repair?

4. How can you make a repair with yourself? What defensive processes may be keeping you unaware of feelings of shame? You may notice repeating experiences in which at first you react intensely to an interaction with your child and only with later reflection become aware of feelings of shame or humiliation. Think of ways in which your own childhood history has given you a sense of disconnection and shame. Let the images and sensations of those experiences come into your awareness. Don't try to censor them; merely observe. When you are ready, ask yourself how you can find a way to heal these old wounds. Let the old issues of the past surface to be recognized and released. You can parent yourself from the inside out.

 SPOTLIGHT ON SCIENCE

**The Tension Between Connection and Autonomy: Complex Systems and Mental Well-Being**

THE COMPLEXITY OF OUR LIVES Numerous investigations into the nature of human experience explore the tension between connection and autonomy. Narratives are filled with the ways the self attempts to find a way to maintain integrity in the face of being connected with others. Developmental studies reveal the ways we struggle between individuation and joining from infancy onward. During adolescence we are faced with the task of finding our own identities away from our parents and seeking novel ways of being in the world. But at the same time, we become more focused on the world of our peers and are influenced by the culture of adolescence that colors the world we live in. As one teen said, "I need to wear these pants like this so I can be like everyone else who is trying to be different." But the tension doesn't end with puberty. Please see Dan's book on adolescence, *Brainstorm,* to see how these issues continue to unfold from the inside out during this important period of life. Adults in their maturity often struggle, perhaps with less turmoil but with similar themes, between commitment to a group and time for autonomy.

Why is there such a tension? We've seen how all children—perhaps all people—oscillate between needs for connection and needs for solitude. Trying to deepen our understanding of this complex issue leads us to the study of complex systems. Known as chaos theory, complexity theory, or the "nonlinear dynamics of complex systems," this view is a mathematically derived perspective on how complex systems such as clouds—and human minds—organize their functioning across time. We'll dive into the waters of complexity briefly through this section and point out some of its principles that are particularly relevant to understanding parenting and relationships.

Why do clouds form in the wondrous ways that they do? Why don't those water molecules that are the components of clouds just randomly distribute themselves throughout the atmosphere? Or why don't they just line up in a row as a single strip of vapor across the sky? Scientists puzzled by these kinds of questions used probability theory to explore some possible answers. Their mathematical formulas suggest a number of reasons why clouds, and other complex systems, flow across time in the trajectories or pathways that they take. A complex system is one that is open (receives ongoing input from outside of itself, such as sunlight, or signals from other people), and has layers of components capable of chaotic behavior. The human mind and brain meet these criteria for complexity.

The basic idea is that complex systems have an innate self-organizing property that is built into the physical relatedness of their components. Self-organization determines the state of the system's flow across time, meaning the position or activity of the system's components. For a cloud, this means where the molecules are and how they are moving at a given moment in time. For a brain, it means which neuronal groups are firing. For a mind, it means how information and energy are flowing. Here are some features of such a self-organizing system:

- Self-organization—The system tends to move toward complexity.

- Nonlinearity—Small changes in the input to the system can lead to large and unpredictable changes in the system's flow over time.

- Recursive features—Over time, the system's flow tends to feed back on itself to reinforce its direction.

- Constraints—Two general factors tend to influence the pathway of the system, internal and external constraints.

- States—The system moves through time by entering states of being, or activation. Certain states are more likely than others to be entered. Constraints, internal and external, can influence the likelihood of certain states being created: (1) attractor states are those with ingrained features making them more likely to occur; (2) repellor states are those whose features make them less likely to occur; (3) the movement between states often involves disorganization and reorganization of the system's flow—these are intermediate, transition states; (4) strange attractor states may occur in which unlikely and usually unstable configurations of the system actually become reinforced and more likely to be activated.

**MENTAL WELL-BEING AND COMPLEXITY** A further application of this view from complexity theory has to do with the notion of mental health and emotional well-being. Complexity theory suggests that the most stable, flexible, and adaptive flow of states occurs when the self-organizing processes of a system move it toward complexity. It's difficult to describe exactly what complexity is, so let's start with the extremes on the scale of degree in order to see what it is not. At one extreme are the conditions of sameness, rigidity, predictability, and total order; on the other end are the conditions of change, randomness, unpredictability, disorder, and chaos. Complexity lies between these extremes.

The example of a choir can help illustrate the sense of complexity. If all the singers sing exactly the same notes in the same way, this would be rigidity—a boring, but loud, output of sounds. If each member of the choir sings totally independently of one another, this would be cacophony and chaos. Complexity, the path between these two extremes, is the same as harmony. Harmony emerges when differentiation and linkage unfold over time. That's our definition of integration. And so complex systems self-organize in an optimal way by linking differentiated elements to each other. The subjective feeling at the extremes away from this integration is boredom on the one hand and anxiety on the other. The rich, vitalizing, energized sense of complexity, of integration, has a vibrant quality to it that emerges when systems are able to move in their natural, self-organizational flow toward complexity, toward harmony.

Not only is such a state full of a sense of life, but complexity theory also predicts that such a self-organizational flow is the most stable, adaptive, and flexible. This is a terrific working definition for mental health! And the notion is that health arises from integration. Please see Dan's *Developing Mind* and *Pocket Guide* if you want to dive into this with more depth!

How does the complex system of the mind achieve such flow? First of all, there is natural movement toward complexity, a natural push of the mind toward health. This is great news! Healing requires the release of that natural, innate process. Experiences and situations that impair the movement of our minds toward complexity can be defined as "stressing" the system. A stressed system is one that tends to move away from harmonious complexity, toward either extreme: rigidity on the one side and chaos on the other.

When we apply this view to an individual mind, we can understand balanced functioning in a new light. When the mind is in a healthy state of flow, energy and information move in ever-evolving ways toward maximizing complexity. Lifelong learning is one example of such a state of continued openness to change. However, sometimes the mind becomes stressed, and the flow of its energy and information moves into "attractor states" that are unhealthy. Sinking into a shame-filled state of withdrawal would be an example of moving away from harmony and entering rigidity. Another example would be becoming infuriated and entering a state of rage in which our minds become chaotic. Both withdrawal and rage are movements of our mind's system away from the harmony of complexity.

**THE COMPLEX TENSION BETWEEN CONNECTION AND AUTONOMY** Building on these principles, we can view the question of the tension between autonomy and connection in a new light. The human mind is certainly an open system (receiving input from outside itself) that is capable of chaotic behavior—too capable for some of us! The mind is a dynamic, complex system impacted by internal and external constraints. In Dan's work in interpersonal neurobiology, the mind is defined as a self-organizing, emergent, embodied and relational process that regulates the flow of energy and information. The short version is simply this: The mind can be defined as an embodied and relational process that regulates the flow of energy and information. The mind is both within us (it is embodied), and it is between us (it is relational). The internal constraints of the mind can be thought of as the synaptic connections among the neurons of the brain, which govern the nature of the flow of energy and information inside the body, which in turn shape mental processes. The external constraints of the system can be seen as the relationships we have with other people. In mind terms, relationships involve the passage of energy and information from one person to another. In other words, interpersonal communication is the external constraint that also shapes the system of the mind. Naturally, we have consciousness and subjective experience, an inner mental life that is filled with feelings, thoughts, and memories. But beyond these aspects of mind, seeing this function of a self-organizing process empowers us to build our regulatory functions. These functions are internal, and they are interpersonal. The mind is within us, and between us.

When we need connection, we are focusing on the use of external constraints to modify the complex system of our own minds. The brain is constructed to rely on external social input to regulate its own functioning. Early in life, infants need connections to caregivers in order to organize their brain's function in the moment, and to allow it to develop properly over time. This is called "dyadic regulation" in that the interactions of the pair or dyad (child and parent) enable the child to achieve balance or regulation within his own mind. Interactions with caregivers allow the child's brain to develop the neural structures necessary to move from dyadic regulation to more autonomous forms of self-regulation. Self-regulatory structures in the brain include the integrative regions of the prefrontal cortex, including the ventrolateral and medial prefrontal regions, and the orbitofrontal and anterior cingulate cortices. These regions take in information from the social world in the form of interpersonal communication. They use this information to help regulate the states of information flow, emotional processing, and bodily balance. The linkage of these four elements of social, cognitive, emotional, and somatic processes is the fundamental integrative and regulatory role of the prefrontal regions.

For example, interpersonal interactions can enable the prefrontal region to modulate the two branches of the autonomic nervous system. The sympathetic (accelerator) and parasympathetic (brakes) branches of this system are under the direct influence of the orbito-frontal cortex, especially in the right hemisphere. The central importance of nonverbal signals from others, perceived and processed by the right hemisphere, can be understood as the ways the right hemisphere uses this form of information to link the interpersonal with the internal. The connection of bodily functions to emotional, cognitive, and social domains reveals the ultimate integrative process of the complex systems of our minds as social and embodied beings.

With adaptive self-regulation, the two branches of the autonomic nervous system are flexibly in balance. Such a situation may be seen as maximizing complexity, and hence the adaptive and stable nature of the state of self-organization. The imbalance of these processes may also reveal movement toward either extreme away from complexity. With excessive sympathetic accelerator activity, the self becomes flooded with activation and enters an overexcited, chaotic state. Such may be the case with agitation and out-of-control anger. With excessive parasympathetic activity, the self becomes shut down and there is a movement into rigidity and emotional paralysis. This condition may be reflected in states of marked despair and depression. A condition may also exist in which both the brakes and the accelerator are engaged simultaneously, as may be revealed in a "strange attractor" state of infantile rage. Movement into these various states of imbalance may be induced by interactions with the social environment and also may be made more likely by experiences in the past that have created vulnerabilities within the individual. These vulnerabilities are embedded in memory, they are our "hot buttons," which directly influence our self-organizational patterns.

External constraints (relationships) enable the young child to develop the emotional regulatory capacity to also rely on internal constraints (neural structure and function that emerges from neural connectivity). In the brain, the balance between these external and internal regulatory processes seems to be mediated by our prefrontal regions. Throughout life we all have cycling needs for connection and solitude. We self-organize by relying on oscillating influences of the external and internal constraints that shape the pathways of the complex systems of our minds.

**INTEGRATION IS BALANCING DIFFERENTIATION AND LINKAGE** How can we conceptualize the path toward complexity in practical terms? Fortunately we can translate the ideas of complexity theory into less abstract, hopefully more accessible concepts. As mentioned briefly above, complexity theory mathematically derives the idea that complexity is achieved when systems are able to balance the two contrasting processes of differentiation (components being specialized) and linkage (components coming together as a functional whole). For example, if the system we are examining is the individual, we can consider how the circuits of the brain are able to become differentiated and then functionally linked in the process of neural integration. The integration of differentiated components enables the system to move toward *maximal complexity. This is the flow of harmony.* We have defined these highly adaptive, flexible, and stable states as synonymous with well-being.

The concept of the complex system can be applied at any number of levels of analysis, including that of the individual, a pair, a family, a school, a community, a culture, and perhaps a global society. Giving respect to the separate and unique identities of individuals in these social groupings enables differentiation to be celebrated. Facilitating the gathering together of these individuals into a functionally linked collaboration permits integration. When linkage and differentiation are in balance, integrative complexity, and therefore well-being, are achieved.

We can use the perspective offered by the lens of complexity theory to understand interpersonal communication. A state of harmonious communication requires that both members be respected for their individuality but come together to integrate their differentiated selves. There is simultaneously a sense of integrity of the individuals and a sense of joining between them. This is the idealized solution to the tension between the drive toward connection and autonomy. Research on securely attached children and their parents reveals evidence of this maximally complex give-and-take in which each person contributes to the dialogue and can anticipate, but never fully predict, the response of the other. There is a vibrant matching, an aliveness, to the connection between the two in such a dyadic form of resonance.

In some families, there may be movement away from such vibrant, complex levels of joining and individuality. At one extreme are "enmeshed" families in which individuality is prohibited. Everyone must like the same food, have similar behaviors, and have the same views. Differentiation is severely impaired, limiting the degree of complexity (and vitality) of the family system. By contrast, in some families linkage is absent. Family members do not share mealtimes, interests, or activities. Conversations reveal a lack of interest in the internal lives of the family members. The highly differentiated individuals are able to carry out their lives without a sense of joining. Without the balance of linkage to counter such differentiation, the level of complexity in this family system is also severely compromised. Both extremes of impaired integration would be examples of stressed systems.

In the day-to-day life of healthy families, the inevitable ruptures in connections among family members may also be understood through the perspective of complexity theory. With rupture, the vibrant sense of being alive and connected is disrupted. The brain may enter states of either chaos or rigidity, and the integration between the two communicators is broken. Sometimes such a rupture can occur when there is a sense of intrusion and pressure to conform to the expectations of the other person. This can be seen as communication that moves toward excessive linkage without respecting the uniqueness of each individual's differentiated self. At the other extreme, disruption may occur when signals are ignored, when linkage or joining is longed for but fails, and the individual is left in a state of excessive differentiation without integration. This isolated state leaves the person to rely on autonomous self-regulation at a moment when connection may have been needed. Either form of disconnection can result in the movement of the individual away from complexity, stressing the system, entering reactive states of rigidity or chaos, and moving away from the balanced capacity for integrative self-organization.

Over time, people develop adaptations, sometimes called patterns of defense, which can be seen as the ways the mind has responded to these moments of intrusion or isolation. When an adult then becomes a parent, these patterns of how the tension between autonomy and connection was handled in earlier family life can become activated in this new setting. People may become lost in familiar places as they reactivate old patterns of defense in response to the intimate connections between parent and child. Finding a way to approach this universal tension, to move between autonomy and connection, between differentiation and linkage, is a lifelong challenge of integration for us all.

## Digging Deeper

L. Chamberlain. "Strange Attractors in Patterns of Family Interaction." In R. Robertson and A. Combs, eds., *Chaos Theory in Psychology and the Life Sciences,* pp. 267–273. Mahwah, N.J.: Erlbaum, 1995.

D. Cicchetti and F. A. Rogosch, eds. *Self-Organization.* Special issue, *Development and Psychopathology* 9, no. 4 (1998).

M. D. Lewis. "Personality Self-Organization: Cascading Constraints on Cognition-Emotion Interaction." In A. Fogel, M. C. D. P. Lyra, and J. Valsiner, eds., *Dynamics and Indeterminism in Developmental and Social Processes,* pp. 193–216. Mahwah, N.J.: Erlbaum, 1997.

A. Schore. "Early Shame Experience and the Development of the Infant Brain." In P. Gilbert and B. Andrews, eds., *Shame: Interpersonal Behaviour, Psychopathology, and Culture,* pp. 57–77. London: Oxford University Press, 1998.

D. J. Siegel. *Mindsight: The New Science of Personal Transformation.* New York: Random House, 2010.

———. *The Developing Mind, Second Edition: How Relationships and the Brain Interact to Shape Who We Are.* New York: Guilford Press, 2012. Chapters 6 and 7.

———. *Pocket Guide to Interpersonal Neurobiology: An Integrative Handbook of the Mind.* New York: W. W. Norton, 2012.

———. *Brainstorm: The Power and Purpose of the Teenage Brain.* New York: Tarcher / Penguin, 2013.

D. J. Siegel and T. Payne Bryson. *The Whole-Brain Child: 12 Revolutionary Strategies to Nurture Your Child's Developing Mind.* New York: Random House, 2011.

L. A. Sroufe. *Emotional Development: The Organization of Emotional Life in the Early Years.* New York: Cambridge University Press, 1996.

L. A. Sroufe, B. Coffino, and E. A. Carlson. "Conceptualizing the Role of Early Experience: Lessons from the Minnesota Longitudinal Study." *Developmental Review* 30, no. 1 (2010): 36–51.

F. Walsh. *Normal Family Processes: Growing Diversity and Complexity,* third edition. New York: Guilford Press, 2012.

# CHAPTER 9

# How We Develop Mindsight: Compassion and Reflective Dialogues

## INTRODUCTION

As parents we offer our children not only experiences that help shape their developing minds but also ourselves. Children learn from us by watching us and modeling what we do. If teaching were telling, the job would be easy. Children will learn what's important to us and what we value by living with us, not just by hearing what we say. Who we are, the nature of our character, is revealed in how we live and how we make decisions about what we do. No matter how much we reflect and deepen our introspection, ultimately how we act in the world gives the true message of our values. Children observe these outward expressions of our character and remember, imitate, and re-create these ways of being in the world. The old adage, "Do what I say, not what I do" is only wishful thinking on the part of parents. Our children watch us because they want to know who we are.

Character develops from how innate constitutional features are molded by the social world of experience. What is the nature of the character we'd like to see our children acquire? If we want our children to grow up to be compassionate, with the ability to respect and care about themselves as well as others and the world around them, we'll want to parent in ways that support the development of empathy. The ways in which we relate to our children can foster their compassion and empathy. Caring individuals who can think and reason, who can enjoy life and build healthy relationships with others, are better able to use their unique gifts as productive members of the community. If helping to develop empathy in our children is the goal, what is known about how it can be promoted by parents?

## INTENTIONALITY AND MINDFULNESS

Being intentional as a parent is a good place to begin. No book or professional can offer the right answers for each possible situation that can arise in the day-to-day life with our children. Instead of depending on techniques alone, parents can learn ways of being with their children that promote the development of presence, kindness, empathy, and compassionate understanding.

This way of being is rooted in a parent's compassionate self-understanding. When we begin to know ourselves in an open and self-supportive way, we take the first step in the process that encourages our children to know themselves. This intentional stance, this attitude of being centered in self-awareness, is a purposeful, mindful approach toward parenting.

How does mindful self-understanding promote empathy? Research in child development and more recently in neurobiology supports the view that three aspects of our brain function coexist and determine how we grow: mindsight (overlapping with processes called "theory of mind," mind-mindedness, reflective function, or "mentalization"), self-knowledge (autobiographical memory and narrative), and response flexibility (executive function, with the ability to plan, organize, and delay gratification). These higher-order mental processes enable us to respond with thoughtful intention in various settings, including difficult and ambiguous situations.

Having well-developed executive functions, having intentional approaches to what we do and the ways that we make decisions, enables us to be flexible in our actions. Executive capacities, combined with a positive concern toward others and an emotional understanding of ourselves, enable us to act compassionately and empathically. Children who develop the interrelated processes of mindsight, self-knowledge, and executive functions are better able to make intentional choices in their behaviors.

Mindful awareness is a formal process, studied in great detail at the beginning of this millennium and practiced in many cultures, east and west, over thousands of years. When we first wrote about being mindful, it was with the general notion of being intentional and kind in mind. As the science of mindfulness has grown, it has become clear that the presence and openness inherent in mindful awareness overlaps directly with an inside-out approach. The acronym COAL is a helpful way to link the two as we are curious, open, accepting, and loving.

What can we do as parents to promote the development of these attributes and abilities in our children? Various studies have shown that parents can actively promote a child's development of the capacity to understand the inner lives of themselves and others with kindness and compassion by engaging in interactions such as pretend play, storytelling, and talking to children about emotions and their impact on how they behave. In many ways, these elements are the basis of social and emotional intelligence, processes that have mindsight at their root.

## MINDSIGHT

Mindsight is the ability to perceive the internal experience of another person and make sense of that imagined experience, enabling us to offer compassionate responses that reflect our understanding and concern. Mindsight also enables us to understand ourselves, as well as to promote integration in our lives. Insight, empathy, and integration are the three pillars of mindsight in our lives. Putting ourselves in another person's shoes requires that we be aware of our own internal experience as we allow ourselves to imagine another's internal world. Such a process creates an image of the mind of another inside of our own. And honoring differences between ourselves and others, and then linking our differentiated selves with compassionate communication, is how mindsight also promotes integration.

The development of mindsight enables children to predict and explain what they see others do in terms of what is going on inside of the others' minds. In developing this process they create a model that includes the sense that others have a mind that is the motivating source of their actions. Understanding the mind of others gives children a way to understand behavior and the social world in which they live. This understanding begins as an infant's ability to recognize the difference between animate and inanimate objects. Children quickly learn the basic principles for human interaction, shaping their expectations for contingency, reciprocity, communication, shared attention, and emotional expression. These interactive experiences shape their ongoing theory or understanding of how human minds work.

Mindsight abilities enable the child to "see" the mind of another person; that is why we use the term "mindsight" for this important human capacity. When we can see the mind of another person, we can understand what that person is thinking and feeling and respond with empathy. Mindsight may enable us to have empathic imagination, whereby we consider the meaning of events not only in our own lives but in others' lives as well. Empathic imagination helps us to understand the intentions of others and to make flexible decisions about our behaviors appropriate to the social situation. Mindsight allows us to understand others and also deepens an understanding of our own mind.

Scientists have proposed that this capacity for knowing minds (see table on next page) in humans is intertwined with the acquisition of language and higher, abstract thinking ability. Language and abstract thinking extend our vision and allow us to create and manipulate mental images beyond the physical world in front of our eyes. A typical infant is born with the genetic, hard-wired capacity for mindsight, but the development of this cognitive ability is shaped by the experiences in a child's life. Mindsight appears to be dependent upon a coherence-making mental process that involves the integrating fibers of our right hemispheres and our prefrontal regions. Future investigations may generate information on exactly how childhood experiences shape the integrative regions of the brain to allow mindsight to develop.

## ELEMENTS OF MIND

Reflective conversations with our children can help to develop our children's mindsight capacities. When we talk with our children, it is helpful if we understand the basic elements of mind that create their internal worlds. These elements include thoughts, feelings, sensations, perceptions, memories, beliefs, attitudes, and intentions. Here we offer a brief overview of some aspects of these elements of mind.

## TABLE 12. KNOWING MINDS

COMPASSION—The ability to feel with another; to be sympathetic, tenderhearted. Compassion is a caring stance toward the distressful emotional experience of another person. Compassion may depend on mirror neuron systems, which evoke an emotional state in us that mirrors that of another person, enabling us to feel another person's pain, not identify with it as ours, and reach out to help that person in soothing their distress.

EMPATHY—Understanding the internal experience of another person; the imaginative projection of one's consciousness into the feelings of another person or object; sympathetic understanding. This is sometimes considered a cognitively complex process that involves mental capacities to imagine the mind of another, but the term "empathy" actually has many definitions and some include emotional resonance and empathic concern, akin to compassion. Empathy may depend on the capacity for mindsight, mediated by the integrative right hemisphere and prefrontal regions of the brain.

MINDSIGHT—The capacity to "see," or image, the mind of oneself or another, enabling an understanding of behavior in terms of mental processes. Other related terms for overlapping processes are "mentalizing," "theory of mind," "mind reading," and "reflective function." Mindsight involves insight, empathy, and integration.

INSIGHT—The power of penetrating observation and discernment, leading to knowledge. When used with the notion of personal introspection, insight implies deep knowledge of the self. Insight by itself does not necessarily involve the ability for empathy or a compassionate approach toward others.

REFLECTIVE DIALOGUES—Conversations with others that reflect on the internal processes of the mind. Reflective dialogues focus on processes such as thoughts, feelings, sensations, perceptions, memories, beliefs, attitudes, and intentions. At a minimum, these dialogues are how we "SIFT" the mind to explore sensations, images, feelings, and thoughts inside of one another.

## Thoughts

Thinking is how we process information in various ways; it is often outside conscious awareness. We may be consciously aware of the outcome of our thinking, our thoughts, by means of words or images. It is important to understand the meaning behind words and images in order to more fully communicate with each other. We have two fundamental modes of thinking—for simplicity, we call them "left mode" and "right mode."

Word-based thoughts are created by the left mode of processing, a linear and logical analytic mode that attempts to make sense of cause-and-effect relationships. Left-mode processing seeks to make assessments of right and wrong and appears to have little tolerance for ambiguity. Contradictory information is not handled well by this mode, and conflicting points of view may be quickly oversimplified in an effort to resolve the problem and come up with a logical solution the left brain considers to be correct. Nonverbal signals and the context of the social world, the domain of the right hemisphere, are often ignored by this left mode of thought.

Keep these limitations of left-mode processing in mind when reflecting on thinking with your child, and yourself. There are many languages of our inner and social worlds that may be expressed in other ways. Right-mode processing is a different type of processing, which may seem relatively quiet yet may be awaiting the chance to be acknowledged and understood. Right-mode thoughts are experienced as a set of images and sensations that are nonlinear and nonlogical. Although these important processes may be difficult to translate into words, they give us direct access to information about how we feel, remember, and create meaning in our lives.

Infants are primarily right brain–dominant, requiring nonverbal communication with parents. Preschoolers have both hemispheres well engaged, but the connecting fibers of the corpus callosum linking the two halves of the brain are still quite immature. During this time a child begins to develop the ability to put words to feelings. Through the elementary school years and beyond, the corpus callosum matures and enables more highly integrative functioning to be achieved. By adolescence, a process of brain reorganization with remodeling takes place that shifts the nature of thinking in profound ways.

### Feelings

Our internal, subjective experience is filled with a continual ebb and flow of the surges of energy and information processing in our minds. Feelings, our awareness of our inner emotions, reveal the conscious sensation of this basic music of the mind. As these primary emotions become more elaborated our minds create a sense of meaning. Meaning and emotion are intimately woven from the same processes. This is why SIFTing the mind for the emotional experience another person is having is so meaningful, as it reveals the meaning within the flow of mind emerging, moment by moment.

Sometimes our primary emotions become elaborated into categorical emotions, including sadness, anger, fear, shame, surprise, joy, and disgust. We can feel these elaborated, intense, and often externally expressed emotions and label them with words. Sometimes, however, an excessive focus on categorical emotions and their word labels alone can get in the way of our seeing the deeper meaning of an experience for ourselves or our children. Images and body-felt sensations can greatly illuminate the deeper meaning within our emotions and our thinking.

Reflective dialogues focus on the elements of mind, including the important dimension of feelings experienced as primary emotions. In this manner, discussing with your children what they are giving their attention to, what they find is important, whether they feel something is "good" or "bad," what they feel like doing, what something means to them, and giving names to the categorical emotions they may be experiencing are all important aspects of talking about feelings.

## Sensations

Before words, we live in a sea of sensations that shapes our internal subjective experiences. Sometimes these sensations are unformed and without definition. In fact, the dominance of the left-mode type of language-based consciousness may make these ambiguous, dynamic, and fluid internal processes seem unimportant and not worthy of attention. However, recent research suggests that these sensations are a vital clue to knowing what has meaning for us. Bodily sensations are a basis for the brain's rational decision-making process and they give vitality to our lives.

Sensations are the core basis of mental life. In our modern world, awareness of sensations is often overlooked. Yet sensations are a rich, untapped source of insight and wisdom. Coherent self-knowledge may depend upon our becoming more sensitive to and aware of our internal states and the sensations that accompany them.

In reflecting on sensations with our children, it is useful to consider how the body feels. We can ask ourselves and our children how we feel in our bodies. How does your stomach feel right now? Is your heart racing? Does your neck feel tight? Focus on these sensations and think about what they mean for you and for your child. You can ask your child, "What do you think your body is trying to tell you right now?" Be open to these nonverbal messages. And we can first describe what we feel, not explain or interpret. Sensations contain a direct pathway to knowing our children and ourselves.

## Perceptions

Each person has his or her own perception of reality. Respecting the individuality of each person's unique point of view is not just a "nice thing to do" but is also a neurologically validated approach. Being open to the uniqueness of each person's experience is not always easy. We often feel that our views are correct while those of others that are different are distorted. It is easy to become attached to the notion that our own perspective is the only way to see things. Doctors who explore the inner life of the patient, even for just a few empathic moments of checking in with how they feel or what they perceive, have patients who recover sooner from a common cold and have immune function that is more robust.

"Metacognition" means thinking about thinking, and it is a process that children learn. An important part of this process, which starts to develop between three and nine years of age, is learning the "appearance/reality distinction." We come to understand that something may appear in a certain way but that this perception may be different from what that something actually is. For example, young children may believe that what they see on television is real, whereas older children understand that film crews can create special effects that look real but are human creations. Another element of metacognition is "representational change," meaning that how one thinks of something at one time may change later on. You can change your mind. "Representational diversity" is the ability to accept the notion that you may see things one way whereas another person sees the same thing in a very different manner. Such a process is revealed when a child can understand, for example, that for one person a roller coaster ride may be experienced as an exciting adventure, whereas for another it may be frightening.

Another element of the important metacognitive ability is an understanding about emotions: emotional metacognition. Children come to learn that emotions influence the ways that they perceive and behave. They also learn that it is possible to have several seemingly conflictual emotions at the same time. Emotional metacognition and the other features of metacognition are all important components of reflective dialogue. They are also a part of being emotionally intelligent. Not surprisingly, this metacognitive development doesn't end in childhood; we are all open to lifelong learning in this important area.

## Memories

Memory is the fundamental way our minds encode an experience, store it, and retrieve it later on to influence our future experiences and actions. Memory has two basic forms. We are born with emerging implicit-memory capacities, enabling us to store behavioral, emotional, perceptual, and bodily memories. Implicit memory also includes the ability to make generalizations across experiences in the mental models that shape our perception of reality.

Explicit memory develops later. Factual memory usually comes on-line after the first birthday, and autobiographical memory after the second. In contrast to implicit memory, explicit memory carries with it an internal sensation that we are recalling something. This explicit form is what we usually think of as "memory."

Talking to your children about what they remember after their experiences is important for a number of reasons. Parents who take part in this "memory talk" actually have children who remember better. In this co-construction of narrative, parents and children together create a story about their daily lives. Both memory talk and co-construction are a joining together in a collaborative effort to focus on memory and weave a story about what is recalled. By linking past, present, and future in a shared construction of a story, we join together in making sense of ourselves over time. This making-sense process is a central mechanism of achieving a coherent internal sense of self.

## Beliefs

Beliefs are at the core of how we come to know ourselves and others. "Beliefs" refers to theories of how the world works. What we believe comes from how our mental models of the world shape our interpretation of our constructed perception of reality. Beliefs can in part be derived from experiences that shaped our nonconscious mental models and thus may exert their effects even without our awareness.

In reflecting on beliefs with your children, keep in mind that even young children have theories about how the world works. It is useful to explore these beliefs by asking some open-ended questions. "What do you think about why that happened?" "What's your idea about how this works?" "How do you think about it differently now?" "Why do you think she was crying at the party?" There are many ways to focus on beliefs. It is especially important to keep an open mind and listen to your child's point of view. Our beliefs are formed from a wide array of what's happened in the past and what is going on in our lives now.

## Attitudes

More transient than a belief or a belief system is the temporary state of mind that can be seen in one's attitude in the moment. This disposition toward perceiving, interpreting, and responding in a particular way shapes all layers of our experiences. An attitude directly affects how we approach a situation. It shapes how we feel toward something and how we behave in the moment. An attitude can directly influence how we interact with others.

It is often helpful to discuss the idea of attitudes and the notion of a "state of mind" directly with a child. For example, if a child is having an emotional meltdown, it is important later on to actually label and offer a name for that general state, with terms that might be created by the child and yourself such as a tantrum, meltdown, emotional tornado, or volcanic eruption. Discuss what your child felt and thought while in that state of mind. This discussion can allow your child to gain insight into the nature of the shifts in our mental states and emotions and see them as temporary changes that can profoundly influence our attitudes toward others.

### Intentions

We have an intentional stance that creates our sense of the desired future and shapes our behavior. However, we may not always be effective at making the outcomes of our behaviors match our intentions. Our actions may sometimes produce one result when we intended to achieve something else. We may have different intentions that are conflictual with each other in their desired outcomes. This is especially true in the complex interactions of parent-child relationships. You may intend for your child to "be happy" but also intend for him to be taught values as you set limits on his requests. You have multiple layers of intentions that may make your own behaviors conflictual and confusing. Discussing the role of intention in an interaction sheds light on what may have been the true motivation behind the actions or words.

Reflecting with your child on the nature of intention can help him or her understand the difference between a desire and an actual outcome. For example, if your child intended for a behavior to result in her making friends with another child but the action was actually too aggressive and pushed the other child away, it may be helpful to explore with your child her intent and possible alternative actions for the future. What was intended by your child to be an act of invitation was interpreted by the other child as an unfriendly behavior. Clarifying how we can intend one thing that is seen differently by others is important in learning to negotiate in the complex social world. When you reflect on these important aspects of the internal and interpersonal worlds with your children, you will help to develop their social skills and emotional competence.

## REFLECTIVE DIALOGUES

When parents have conversations with their children in which they reflect on the internal processes of people's minds, children begin to develop mindsight. If parents focus only on their children's behavior and do not consider the mental processes that motivate that behavior, they often end up parenting for short-term results and do not help their children learn about themselves. They react quickly and do whatever will make their children or themselves less distressed in the moment. We call this "any port in a storm" kind of parenting. When we seek the closest harbor whenever things get rough, we'll most likely not reach our destination. As parents we have the opportunity to take a "mindsight stance" in our ways of being with our children and help them develop mindsight skills that will enhance their lives for years to come. Thinking about what is important for the long-term development of your children's character can help you be more intentional when you make choices about how to respond to them.

An attitude of love and patience, for our children and ourselves, permits dialogues that value and enhance each person's individuality. We promote mindsight in our children by having a way of being as a parent that respects each person's subjective reality. One way of expressing this intention is in reflective dialogues that engage us in conversations with our children about the nature of our inner mental lives.

For example, when reading a story to your child, you can elaborate the story by discussing what the characters might be thinking and feeling. These kinds of discussions help develop children's empathic imaginations and provide them with the vocabulary to articulate the inner workings of mental life. Language enables us to represent, process, and communicate ideas that advance our abilities to imagine new possibilities for how the social world works. Mindsight abilities expand our capacity for negotiating in a complex world of social interactions. How we use language with our children creates a new level of meaning, a new dimension for how they come to understand their experiences.

One study compared the mindsight abilities of deaf children raised by parents who were fluent with those whose parents were not fluent in the use of sign language. Mindsight was normal in children of fluent signers, but very deficient in those of the nonfluent signers. The cause of this marked difference, researchers have proposed, was that the parents not fluent in sign language did not easily carry out parent-child communication about the mind, whereas these kinds of communication did occur between fluent-signing parents and their children. In other studies, children whose parents talked about emotions, especially focusing on their causes, had a more well-developed understanding of the role emotions play in people's lives. Such conversations, along with pretend play and elaborative storytelling, have all been identified as building blocks of mindsight.

Another research finding is that in families where there was a high degree of parental attempts to control children's behaviors or a lot of negative emotion, the children had impaired mindsight abilities. Other studies have shown that when parents behave in intrusive or frightening ways, their children also tend to have underdeveloped reflective abilities. Researchers have found that intense emotion by itself is not the problem. When parents provide support during difficult emotional experiences, children actually have the opportunity to develop a more sophisticated understanding of the mind. Instead of considering day-to-day friction as only problematic, parents can use these heightened moments of emotional tension to enhance their children's learning through reflective dialogues that deepen their mindsight abilities.

Mindsight is built on word-based reflective dialogues but acknowledges the central importance of respecting the nonverbal, the visceral, and the nonconscious aspects of our lives. Language is valuable because it enables the mind to develop abstract notions about the world. But word-based language is only a part of the picture: studies suggest that mindsight abilities depend in large part on the nonverbal processing of the right hemisphere. In fact, mindsight requires that we have fluid ways of integrating the often ambiguous and subtly unique aspects of social interactions, a right-mode specialty. Creating those representations and holding them in mind long enough to process them are complex abilities. Mentalizing emerges from a "central coherence" integrative capacity that automatically and outside conscious awareness comes up with a sense of the social world.

For example, if you observe your child acting irritably and moping around the house after school, you might have the thought "She must be acting this way now because rehearsals for the school play started today and she was so disappointed that she didn't get a part in it." We come up with a notion based on our observations of a person's present expressions and events that we remember occurred in the recent past. Sometimes we just have a gut feeling about something, and we can state in words our conclusions even though we don't really know how we know. The internal sensations of the body, remember, are represented primarily in right-hemisphere processing, from which so much of our personal self and social minds emanates. Paying attention to our visceral sensations can help us be open to the subjective experience of others and allows you to respond to their minds, not merely their behaviors.

Instead of chastising your daughter because she is irritable and difficult, you might talk with her about how she feels and what may be the cause of her irritable mood. "Did you have a really rough day? Do you want to talk about it?"

Another way to help your child to build his self-understanding is through storytelling. Telling the story of your child's day can help a young child remember and integrate the events of his life. The day's experiences can be revisited in a warm and nonjudgmental way to include the difficult moments as well as the enjoyable ones. In this way the emotional highs and lows of a normal day can be integrated in his memory.

Bedtime can be a good time for reviewing the day's activities. At the end of the day, you might tell the story of your young child's day something like this. Encourage your child to offer his own memories and ideas during the story and to ask questions at any time. "You did so many things today. After breakfast you went to the park and really liked it when I pushed you on the swing. You went so high and stretched your feet out and pushed them back and started to pump all by yourself. How did that feel? You didn't like it when I said it was time to take a nap. When you got up from your nap, I was outside planting flowers in the garden and you were mad because I started working in the garden without you so you pulled up some flowers. I got very angry and used a loud voice and told you to stop doing that right now. Was that scary? You had some big feelings and you cried. After you felt better you planted some red flowers right next to mine. Before dinner you helped me to wash the vegetables for the salad and tore them into little pieces and put them in the bowl. Do you remember what Daddy said? Daddy said that it was a tasty salad. And you seemed very proud of yourself. What else do you remember about today?"

Difficult emotional issues may be best explored during the daylight hours when both child and parent are alert and rested. Telling stories, with drawing or using puppets and dolls, can help a child to understand what happened in order to process and integrate a difficult experience. Telling the story of an experience can help a child become free of confusing or upsetting aspects of a stressful event.

In some ways the parent is the scribe recording the child's experiences and reflecting them back so that the child can make sense of what he is experiencing. It is through these early parental reflections that a child first learns who he is and how to make sense of the world. Reflective dialogues nurture a sense of coherence in that they help to make sense of the internal processes that underlie external behaviors.

## CREATING A CULTURE OF COMPASSION

Reflective dialogues build mindsight abilities by creating a culture of compassion within the family. Culture involves a set of assumptions, values, expectations, and beliefs that shape how we interact with others and influence what has meaning in our lives. In the larger society, cultural practices are at work at many levels of our day-to-day life. Values that emphasize spirituality, altruism, education, materialism, or competition infuse the many different worlds in which we live. In the home, we create a world in which values are expressed either directly or indirectly in our language, in how we act, and in the emphasis we place on the various dimensions of our children's lives. A culture of compassion promotes appreciation of differences, mutual respect, compassionate interactions, and empathic understanding among family members. With thoughtful intent we can choose the values we care about to create a culture in our homes that provides meaning to the daily lives of our children.

A culture of compassion in the home promotes the attitude that sharing others' emotions, feeling their pain and their joys, is valuable. Being empathic extends this emotional resonance into a more conceptual realm, where language can be used to deepen the dialogue and promote understanding. When family members understand and respect each other, they are more likely to act in caring ways. Empathic concern and compassionate behavior can be valued in the home as much as accomplishments in the classroom or on the athletic field. Compassion and kindness are natural outcomes of integration, of honoring differences and linking one another through caring communication. By enacting the values of empathy and compassion ourselves as parents, we can model them for our children.

With mindsight, the understanding of each family member's inner world is given meaning and voice. Each member can differentiate themselves and then link in compassionate connections. Insight, empathy, and integration are the three pillars of a mindsight approach to parenting. Family conversations can weave stories about our lives, using reflective language that embraces the thoughts and feelings of our shared human experience. As mindsight is made visible through reflective dialogues, a common language of the internal becomes a part of family life. The rush of busy schedules may make such a process seem difficult to accomplish, but taking a mindsight stance, embracing a compassionate way of being as a parent, can make it more likely that these important connecting communications will occur. Engaging children in discussions that focus on the subjective experience of others and stimulating their empathic imaginations through pretend play and storytelling enables them to make the inside visible, to bring the inside out.

By developing the capacity for reflective dialogue, we can enrich mindsight capacities in ourselves and in our loved ones. Mindsight is an ability that may continue to develop throughout our lives as we find new and deeper connections to others. By developing reflective dialogues as a part of our lives with our children, we can connect with them in the important process of developing their own mindsight abilities and sense of closeness to us. These connections enable us to transcend the boundaries of our own skins and become a part of a "we" that enriches our lives and perhaps the larger world in which we live.

**INSIDE-OUT EXERCISES**

1. Have a family meeting so that you can get to know what each family member thinks and feels about a particular issue. This is an opportunity to pose questions that encourage the exploration of elements of mind and help develop your children's mindsight. Topics might include things people like or don't like about being in the family or how people feel about some recent family events. Choose a time to meet when parents and children are rested and focused, and you can be uninterrupted. Although each

family member should attend, not everyone may choose to speak. This practice develops respect for both speaking and listening. It can be helpful to designate whose turn it is to speak, and for others to know it is their time to listen. Some families prefer to use a small object held by the speaker, while others simply like to take turns. Taking turns and respecting each other's right to speak and be heard with respect and without interruption builds the value of reflection in the family.

2. With your child, sit down and discuss a shared experience. Notice what different aspects each one of you remembers. Take your child's lead and reflect back elements of each of your experiences of the events. Try to incorporate into the talk a focus on the elements of mind, not merely the external aspects of the experience. What matters most is not to establish the accuracy of the content, but to join together in the telling of the story. The process is the point in the co-constructive experience. Have fun!

3. Create a family book together. Include a chapter for each member of the family to create a story in images and words about themselves as individuals. Entries can include photographs, drawings, stories, and poems, offering an opportunity for the expression of creativity. Other pages can focus on family experiences, such as traditions, celebrations, outings, holiday events, and other people important to the family. This co-constructed story documents the ongoing life of your family and can deepen your connections with each other.

## Language, Culture, and Mindsight

Human beings are social creatures. We have evolved over millions of years of biological time, riding the wave of generations of species before us that have survived through the generation of diversity by means of spontaneous mutations that has enabled us as animals to adapt and reproduce. More recently, our evolution has been shaped not so much by changes in our genetic information and bodily structure as by changes in the ways in which we use our minds. This mental transformation, or cultural evolution, involves the creation of meaning and the sharing of knowledge from one mind to another, and across the generations.

Most mammals are social creatures, a feature mediated by the limbic portions of their brains, and primates are especially able to mirror each other's states in complex forms of communication that enable intricate social functioning. Humans, evolving from a common ancestor we shared with the chimpanzee over five million years ago, have built new mental capacities that have extended beyond what our primate cousins are capable of: we can represent the minds of others, and of ourselves. A number of researchers suggest that this ability to represent minds, called "theory of mind," mentalizing, or mindsight, has enabled us also to develop general theories about the world at large. This theorizing capacity, made possible because of the development in large part of the integrative frontal regions of our cortex, then enables us to develop the cultural components of human life: representational art, representational ideas about the world (science), and representational language (enabling us to write and communicate abstract ideas to others).

The development of this human ability thrust cultural evolution into fast forward. Here are some thoughts of the researcher Steve Mithen about culture and the understanding of minds (Baron-Cohen, Tager-Flusberg, and Cohen 2000, 490; all sources cited below with a page number only are in this anthology).

*The cultural environment must be acknowledged as being an influence over the specific timing and nature of theory of mind development. A theory of mind would substantially enhance the extent, detail, and accuracy of the inferences that could be made. Consequently individuals who had a theory of mind encoded in their genes—in either a strong or a weak sense—would have had considerable reproductive advantage within the societies of our early ancestors. . . . One possibility is that this reflects the evolutionary roots of language as a means of maintaining social cohesion.*

Language is thus seen as evolving as a social function. How then do the kinds of language experiences children have influence their development? Mithen goes on to address this issue:

*By the age of four, human children certainly engage in mind reading; indeed they appear to do it continually and compulsively. Without doubt, a profound difference between the human and chimpanzee. . . . The presence of imaginary beasts in Upper Paleolithic art, together with associated evidence for ritualistic activity—an activity that normally involves pretence—appears conclusive evidence that those cave painters were also mind readers. . . . Some form of linguistic ability appears likely to be a precondition for a theory of mind, either in development or in evolution. (494–496)*

This perspective on evolution and culture enables us to see the link between language development and the capacity to "see" or "conceive" of the mind itself. What happens when a culture's language does not include words about the mind? Such "mental-state" terms are found in so-called Western cultures, but they do not exist in all cultures.

The cultural researchers Penelope Vinden and Janet Astington note that certain factors have been found to be associated with theory-of-mind development, including language development, emotion understanding, pretend play, parenting style, and social class. These investigators raise important general points about the need to have a cultural context in which to understand the research on theory of mind (509–510):

*Language is fundamental because it is through language that we create culture. It is a tool for interacting—language use is not an individual process, but a joint action among participants based on a common ground, which in broadest terms is a shared cultural background. . . . We take a mentalistic stance to ourselves and other people, including the youngest of other people, that is, infants. We ascribe mental states to them in the ways we interact with them and in the lexical terms we use. . . . Thus, in acquiring our language children acquire our theory of mind, which is embedded in speech practices.*

In this manner, the very ways in which we engage with others in communication shape the nature of the reality that we mutually create. For some cultures, creating a shared sense of the existence of mind may be inherent to the assumptions and beliefs of that social system. For other cultures, such a belief may not be built into the daily lives of its members. Vinden and Astington note that not all cultures conceive of mind as we do and they may have quite different conceptions of the mind from us. For other groups, the concept of "mind" may not exist at all. There may be differing approaches to explaining behavior that facilitate social interaction without relying on the concept of mind.

Such cultures may have a lack of mental-state terms and individuals may perform poorly on theory-of-mind tests developed in the West to assess mentalizing abilities. Vinden and Astington go on to state (510): "Perhaps, with our focus on the understanding of mind, we reveal a cultural 'obsession' with minds that may not be universal. Perhaps children elsewhere develop a theory of behavior, or a theory of body, or a theory of possession, or multiple theories that are applied in different culturally defined contexts."

This is a useful perspective as we seek to apply suggestions for developing mindsight. Though all cultures appear to have contingency as a form of essential communication, not all cultures have sophisticated notions of mind, or at least words to express such notions. Whereas the genetic capacity for mindsight appears to be a common human potential—and clear-cut studies point to the neural structures that mediate these functions—perhaps not all cultures build on this innate ability. Shared experiences among individuals in the creation of meaning and defining what is real are essential features of culture. Cultural biases permeate even our basic definitions of mind and of self, and keeping these issues in mind is essential as we interpret research findings and assess the usefulness of parenting suggestions in different cultural settings.

* * *

## A Theory of Theory of Mind: Experiential Building Blocks of Mindsight

Researchers in fields ranging from anthropology to neuroscience are actively exploring the origins of our human ability to understand each other. Wellman and Lagattuta state the fundamental approach (21):

> Humans are social creatures: we raise and are raised by others, live in family groups, co-operate, compete, and communicate. We not only live socially, we think socially. In particular, we develop numerous conceptions about people, about relationships, about groups, social institutions, conventions, manners and morals. The claim behind research on "theory of mind" is that certain core understandings organize and enable this array of developing social perceptions, conceptions and beliefs. In particular, the claim is that everyday understanding of persons is fundamentally mentalistic; we think of people in terms of their mental states—their beliefs, desires, hopes, goals, and inner feelings. Consequently, an everyday mentalism is ubiquitous and crucial to our understanding of the social world.

Simon Baron-Cohen, Helen Tager-Flusberg, and Donald Cohen help to define the importance of theory-of-mind abilities (vii): "Understanding the actions of others and of one's self in terms of agency and intentional states is central to a child's socialization and to the capacity of adults to empathically and correctly understand each other. The capacity to read the minds of others rests upon a long evolutionary history, is grounded in neurobiology, and becomes expressed in the first intimate social relations."

Experiences that promote self-understanding are based on mindsight-building experiences. Understanding one's own mind appears to enhance the capacity to understand others' minds. In turn, parents who communicate with children about emotion, especially using discussions about the causal impact of emotions on behaviors and thinking, have children who understand emotion in a deeper way—in themselves and in others. How are self-understanding and other-understanding related? Chris Frith and Uta Frith explore these issues by examining some of the neurophysiological mechanisms underlying theory of mind (351–352):

> We speculate that one aspect of the human ability to make inferences about the intentions of others evolved from a system concerned with analyzing the movements of other creatures. This information about the behavior of others is combined with information about own mental states represented in medial prefrontal areas. We propose that the critical mechanism that allows humans to develop a theory of own and other minds results from the development of introspection (monitoring one's own mind) and an adaptation of a much older social brain concerned with monitoring the behavior of others.

In this manner, we link our internal experience to our perception of others' behavior.

Human development is thrust forward by social experiences. Our social worlds are not just coincidental environments in which we live, they are the essential social matrix in which our minds have evolved and in which a child's mind develops.

Rhiannon Corcoran examines the way the social cognitive ability of understanding minds may be impaired in a number of situations. She has proposed a model for how theory of mind is built upon experience (403): "The model runs as follows: when people attempt to figure out what another person is thinking, intending, or believing, the initial step is to introspect. We try to determine what we ourselves would think, believe, or intend within the present context. We do this by referring to the contents of our autobiographical memory and any relevant information about ourselves will be retrieved from this source."

This suggests that introspection (self-understanding) would go hand in hand with understanding others and the mind. Corcoran notes that developmental researchers "stress the proximity of emergence of the first theory of mind skills to that of the first autobiographical retrieval skills" (405); see also Howe and Courage 1997; Welch and Melissa 1997. Others have noted the overlap of the processes of mindsight, executive skills, and autobiographical memory, all dependent upon intact right prefrontal functioning. Executive functions include planning, organization, and impulse inhibition and thus are a part of the capacity for response flexibility.

If mindsight skills are entwined with these important mental processes, do we know what experiences can promote their development? Corcoran states (415):

*One of the strongest suggestions made in a number of studies reviewed here is that environmental factors have a strong*

*bearing on the adequate functioning of theory of mind. Paren-*
*tal neglect and abuse and negative or bizarre recollections*
*are all put forward by several authors as possible causes of*
*deficient mentalizing. . . . One of the things that the studies of*
*other clinical samples [besides those of children with autism*
*who appear to have a constitutional neural impairment of the-*
*ory of mind abilities] has pointed out quite clearly is that men-*
*talizing skills are also acquired as a result of interpersonal*
*interactions within the family setting.*

The essential early elements that reveal the emergence of theory of mind include empathic interactions (sharing emotional states),
communicating with reciprocal nonverbal signals (enabling the give-
and-take of communication to occur), joint attention (attending to a
third object together), and pretend play (creating imagined situations
by investing inanimate objects with animate qualities). As the child
develops, mindsight abilities become more evident in examples of
metacognition, or thinking about thinking. Metacognitive development makes children's theories of mind more sophisticated and
adaptive to the complex social world in which they live.

## Mindsight and the Brain: The Integrative Role of the Right Hemisphere and the Prefrontal Regions

How does the brain create a representation of a mind? To understand how we understand the minds of others and ourselves, we can
look to fundamental ways that the brain processes information in
general. The term "central coherence" refers to Uta Frith's notion of
how widely distributed processing circuits become linked into a

coherent whole, proposed to be essential for representing other minds. Francesca Happé explores this idea (215):

*Social understanding cannot be considered independent of coherence—because in order to appreciate people's thoughts and feelings in real life one needs to take into account context and to integrate diverse information. So when we measure social understanding in a more naturalistic or context-sensitive way, we are likely to find a contribution from central coherence—and that individuals with weak central coherence and detail-focused processing are less successful in putting together the information necessary for sensitive social inference.*

How can the mind achieve such an integrative function? The circuits of the brain provide some clues: researchers have found the right hemisphere and the highly integrative prefrontal regions to be active in mentalizing processes. Theory of mind may depend on representing in the mind ambiguous representations or interpretations that are not easily defined, which appears to be a right-hemisphere specialization (Brownell et al., 320–323):

*The right hemisphere is more likely than the left to process weak or diffuse association and low frequency alternative meanings, and the right hemisphere maintains activation over longer prime-target intervals. Whenever there is not a single appropriate, highly activated interpretation, the right hemisphere will play a larger role: when several considerations*

*must be integrated or when an initially attractive interpreta-*
*tion must be abandoned in favor of another. . . . Right*
*hemisphere is necessary for the activation of extensive repre-*
*sentational sets and, in this vein, for meaning that is implied*
*or novel rather than explicit and based on existing rou-*
*tines. . . . Right hemisphere shows superiority in the realm of*
*implied meaning.*

The authors note that it is not always possible to draw a clear differentiation between the functions of the prefrontal regions and those of the right hemisphere as revealed in scanning studies. In this manner, both right hemisphere and prefrontal areas have been implicated in theory-of-mind tasks. Both areas of the brain are highly involved in integrative functions.

To accomplish the integrative functions necessary for theory-of-mind processes, the brain links the two hemispheres to the prefrontal regions. In order to achieve a mentalizing function, these integrative circuits must maintain representations and process various uncertain combinations (323).

*The selection of an appropriate response to an ambiguous*
*or multistep task requires the maintenance of heteroge-*
*neous representational sets for an extended period of time.*
*The high degree of connectivity between the prefrontal*
*regions to other, more posterior and subcortical, brain re-*
*gions makes this region architectonically well-suited for the*
*maintenance of the extensive network necessary for such*
*computations.*

Depending on the nature of the social interpretative task, different areas of the brain may become engaged. The neural processing underlying the inferences of the content of mental states, the researchers propose, will vary depending on the task. More novel tasks that require maintaining and updating ambiguous and relational data will require right-hemisphere activation, while tasks with more routine, scriptlike demands may be processed largely by the left hemisphere circuits.

Summarizing their view of the research literature, Brownell and colleagues state (326–327):

> We argue that the right hemisphere and prefrontal regions both are required for normal ToM [theory of mind]; impairment in either can disrupt task performance. The need to maintain alternative readings of tangential associations for longer periods of time, the novelty of the situation (that is, the absence of an appropriate decision-making algorithm), and the affective marking of an alternative are all features that will increase the potential contribution of the right hemisphere. . . . In general, the right hemisphere helps preserve the raw material from which the prefrontal and limbic regions draw in decision-making processes.

One can see from these discussions that mentalizing requires the coordination of various neural circuits that enable a highly integrated form of social cognition to occur. As these circuits develop in the child, we can imagine that the experiences in families that promote their integrative functioning would be those that involve reflective dialogues and other interpersonal communication that deepens children's understanding of others, and of themselves. Future interdisciplinary investigations may clarify exactly how experiences within families promote the integrative prefrontal and right-hemisphere processes that enable the human capacity for mindsight to develop.

## Digging Deeper

A. Abu-Akel and S. Shamay-Tsoory. "Neuroanatomical and Neurochemical Bases of Theory of Mind." *Neuropsychologia* 49, no. 11 (2011): 2971–2984.

R. Adolphs. "The Neurobiology of Social Cognition." *Current Opinion in Neurobiology* 11 (2001): 231–239.

S. Baron-Cohen, H. Tager-Flusberg, and D. J. Cohen, eds. *Understanding Other Minds: Perspectives from Developmental Cognitive Neuroscience.* Oxford, U.K.: Oxford University Press, 2000. Several articles in this edited volume are referenced in this chapter's "Spotlight on Science."

N. Benbassat and B. Priel. "Parenting and Adolescent Adjustment: The Role of Parental Reflective Function." *Journal of Adolescence* 35, no. 1 (2012): 163–174.

H. Brownell et al. "Cerebral Lateralization and Theory of Mind." In S. Baron-Cohen, H. Tager-Flusberg, and D. J. Cohen, eds., *Understanding Other Minds,* pp. 306–331. Oxford, U.K.: Oxford University Press, 2000.

R. Coles. *The Moral Intelligence of Children.* New York: Penguin/Putnam, 1998.

———. *Lives We Carry with Us: Profiles of Moral Courage.* New York: The New Press, 2010.

R. Corcoran. "Theory of Mind in Other Clinical Conditions: Is a Selective 'Theory of Mind' Deficit Exclusive to Autism?" In S. Baron-Cohen, H. Tager-Flusberg, and D. J. Cohen, eds., *Understanding Other Minds,* pp. 391–421. Oxford, U.K.: Oxford University Press, 2000.

M. J. Doherty. *Theory of Mind: How Children Understand Others' Thoughts and Feelings.* New York: Psychology Press, 2009.

J. Dunn et al. "Young Children's Understanding of Other People's Feelings and Beliefs: Individual Differences and Their Antecedents." *Child Development* 62 (1991): 1352–1366.

R. Ensor, D. Spencer, and C. Hughes. "'You Feel Sad?' Emotion Understanding Mediates Effects of Verbal Ability and Mother–Child Mutuality on Prosocial Behaviors: Findings from 2 Years to 4 Years." *Social Development* 20, no. 1 (2011): 93–110.

P. Fonagy et al. "The Capacity for Understanding Mental States: The Reflective Self in Parent and Child and Its Significance for Security of Attachment." *Infant Mental Health Journal* 12 (1991): 201–218.

P. Fonagy and M. Target. "Attachment and Reflective Function: Their Role in Self-Organization." *Development and Psychopathology* 9 (1997): 679–700.

C. Frith and U. Frith. "The Physiological Basis of Theory of Mind: Functional Neuroimaging Studies." In S. Baron-Cohen, H. Tager-Flusberg, and D. J. Cohen, eds., *Understanding Other Minds,* pp. 334–356. Oxford, U.K.: Oxford University Press, 2000.

F. Happé. "Parts and Wholes, Meaning and Minds: Central Coherence and Its Relation to Theory of Mind." In S. Baron-Cohen, H. Tager-Flusberg, and D. J. Cohen, eds., *Understanding Other Minds,* pp. 203–221. Oxford, U.K.: Oxford University Press, 2000.

M. L. Howe and M. L. Courage. "The Emergence and Early Development of Autobiographical Memory." *Psychological Review* 104 (1997): 499–523.

M. H. Johnson and M. De Han. *Developmental Cognitive Neuroscience,* third edition. Sussex: Wiley-Blackwell, 2011.

F. C. Kobayashi and E. Temple. "Cultural Effects on the Neural Basis of Theory of Mind." *Progress in Brain Research* 178 (2009): 213–223.

J. D. Lane, H. M. Wellman, S. L. Olson, J. LaBounty, and D. C. Kerr. "Theory of Mind and Emotion Understanding Predict Moral Development in Early Childhood." *British Journal of Developmental Psychology* 28, no. 4 (2010): 871–889.

J. Laranjo, E. Meins, and S. M. Carlson. "Early Manifestations of Children's Theory of Mind: The Roles of Maternal Mind-Mindedness and Infant Security of Attachment." *Infancy* 15, no. 3 (2010): 300–323.

S. Mithen. "Paleoanthropological Perspectives on the Theory of Mind." In S. Baron-Cohen, H. Tager-Flusberg, and D. J. Cohen, eds., *Understanding Other Minds,* pp. 488–502. Oxford, U.K.: Oxford University Press, 2000.

C. C. Peterson and M. Siegal. "Domain Specificity and Everyday Biological, Physical, and Psychological Thinking in Normal, Autistic, and Deaf Children." *New Directions for Child Development* 75 (1997): 55–70.

C. C. Peterson, H. M. Wellman, and V. Slaughter. "The Mind Behind the Message: Advancing Theory-of-Mind Scales for Typically Developing Children, and Those with Deafness, Autism, or Asperger Syndrome." *Child Development* 83, no. 2 (2012): 469–485.

A. Shahaeian, C .C. Peterson, V. Slaughter, and H. M. Wellman. "Culture and the Sequence of Steps in Theory of Mind Development." *Developmental Psychology* 47, no. 5 (2011): 1239–1247.

D. J. Siegel. *The Developing Mind, Second Edition: How Relationships and the Brain Interact to Shape Who We Are.* New York: Guilford Press, 2012. Chapter 9.

———. *The Mindful Brain.* New York: W. W. Norton, 2007.

———. *Mindsight.* New York: Random House, 2010.

———. *Brainstorm.* New York: Tarcher/Penguin, 2014.

S. Taubner, L. O. White, J. Zimmermann, P. Fonagy, and T. Nolte. "Attachment-Related Mentalization Moderates the Relationship Between Psychopathic Traits and Proactive Aggression in Adolescence." *Journal of Abnormal Child Psychology* 41 (2013): 1–10.

M. Tomasello. *The Cultural Origins of Human Cognition.* Cambridge, Mass.: Harvard University Press, 1999.

———. "Biological, Cultural and Ontogenetic Processes." In *Constructing a Language: A Usage-Based Theory of Language Acquisition*, chapter 8. Cambridge, Mass.: Harvard University Press, 2005.

P. G. Vinden and J. W. Astington. "Culture and Understanding Other Minds." In S. Baron-Cohen, H. Tager-Flusberg, and D. J. Cohen, eds., *Understanding Other Minds,* pp. 503–520. Oxford, U.K.: Oxford University Press, 2000.

E. A. Von dem Hagen, R. S. Stoyanova, J. B. Rowe, S. Baron-Cohen, and A. J. Calder. "Direct Gaze Elicits Atypical Activation of the Theory-of-Mind Network in Autism Spectrum Conditions." *Cerebral Cortex* (2013); first published online January 16, 2013.

R. Welch and K. Melissa. "Mother-Child Participation in Conversation About the Past: Relationship to Preschoolers' Theory of Mind." *Developmental Psychology* 33 (1997): 618–629.

H. M. Wellman and K. H. Lagattuta. "Developing Understandings of Mind." In S. Baron-Cohen, H. Tager-Flusberg, and D. J. Cohen, eds., *Understanding Other Minds,* pp. 21–49. Oxford, U.K.: Oxford University Press, 2000.

H. M. Wellman, F. Fang, and C. C. Peterson. "Sequential Progressions in a Theory-of-Mind Scale: Longitudinal Perspectives." *Child Development* 82, no. 3 (2011): 780–792.

# Reflections

Parenting is an opportunity for lifelong learning. Children offer us the chance to be in a relationship that encourages us to deepen our connections with others and with ourselves.

We want to be the best parent we can be. Even if our own experiences did not provide us with the childhood we would want for our own children, we are not destined to repeat the past. By making sense of our own childhoods, we have the opportunity to bring openness to our experiences and choice to our daily interactions with our children. When we become free from the baggage of the past, we can live more fully with a flexible sense of fresh anticipation and spontaneity.

Unresolved and leftover issues can impair our abilities to offer our children joyful connections and secure attachments. With security, children have the foundation for healthy development. The great news from research is that an individual's attachment can move toward security with the emotional connections of collaborative relationships. As parents we want to provide the essential ingredients of secure attachments as early as possible, but it's never too late to begin. As we learn to apply the processes of contingent communication, response flexibility, rupture and repair, emotional connection, and reflective dialogues to our interactions with our children, we promote their security of attachment.

We don't need to have had great parents in order to parent our own children well. Being a parent gives us the opportunity to reparent ourselves by making sense of our own early experiences. Our children are not the only ones who will benefit from this making-sense process:

we ourselves will come to live a more vital and enriched life because we have integrated our past experiences into a coherent ongoing life story.

The amazing finding that the most powerful predictor of a child's attachment is the coherence of the parent's life narrative allows us to understand how to strengthen our children's attachment to us. We are not destined to repeat the patterns of the past because we can earn our security as an adult by making sense of our life experiences. In this way, those of us who have had difficult early life experiences can create coherence by making sense of the past and understanding its impact on the present and how it shapes our interactions with our children. Making sense of our life stories enables us to have deeper connections with our children, and to live a more joyful and coherent life.

Integration may be at the heart of living a coherent life. Being mindful, having parental presence, living in the moment, and being open and receptive to our own and others' experiences enables us to begin a process of deepening self-awareness. When we speak of integration, we are talking about a process that links the here and now with the transcendence of time. Integration also involves connecting our emotions and bodily sensations as we create the ongoing story of our lives within our thoughts and through our actions.

Because emotion is an integrative process, how we come to balance and share our emotions reflects the ways we achieve integration within ourselves and with others. As we become connected to our own feelings, we are ready to connect with others. It is through the sharing of feelings that we create meaningful connections. With integration, coherence is created within the mind that nurtures a deep sense of vitality, connection, and meaning.

In the high mode of processing, we can pause, reflect, consider various options of response, and then choose from a wide range of possibilities. Emotional intelligence is synonymous with insight, empathy, and flexibility. When our minds enter the low road, we shut off these integrative capacities and enter the low mode of processing: automatic knee-jerk reactions dominate our responses. Our unresolved and leftover issues hold sway. We become stuck.

Each of us may have leftover and unresolved issues in need of careful mindfulness. Each of us at various times can be bound for the low road. If we let pride, or shame, block us from recognizing these patterns, we will lose the mindsight abilities that can free us from the prison of the past.

Mindsight depends upon a form of integration that enables us to see the minds of others as well as of ourselves. Mindsight allows us to focus on the elements of mind—thoughts, feelings, perceptions, sensations, memories, beliefs, attitudes, and intentions—that form the core of our subjective internal worlds.

The essence of living a coherent life is in the integration of various domains of experience. Integrating our bodily sensations, our emotions, and our life histories is at the heart of mindsight and mindful living. This enhanced self-knowledge allows us to create a sense of connection to ourselves and to our children as we link the present to the past, and become active authors of our ongoing life stories.

The process of emotional attunement enables us to achieve a direct connection between our children and ourselves. This alignment is a form of interpersonal integration. At the heart of this attuning is the sharing of nonverbal signals, including tone of voice, eye contact, facial expressions, gestures and touch, and timing and intensity of responses. Being mindful of these signals in our children is paralleled by our own awareness of the sensations of our own bodies. Bodily sensations form an important foundation for knowing how we feel and what has meaning in our lives. Emotional communication enables us to viscerally feel the joy of our children and to share and amplify such positive states with them. Such emotional joining also enables us to feel our children's pain and then to soothe their distress with our nurturing connection. This is the basis of love. Emotional communication connects us more fully to others and to ourselves.

As writers of this book, we have loved working together to try to put words to these processes. We have drawn on what we have learned from ourselves, from our work, and from our studies. The work of trying to create coherence is the work of life. In its deepest form it is the integration of energy, of information, of the essence of mind. We engage in mental time travel, connecting past, present, and future, as we make sense of our lives and deepen an integrated self-knowledge. Though words are truly limited in their ability to describe this process, it boils down to this: we connect. We connect to each other across the boundaries of time and space that separate one mind from another. We connect to our own life stories as they unfold across the life span.

Creating coherence can be a lifelong adventure. Integrating self-understanding is a challenge that never ends. Making that challenge a journey of discovery comes from openness to growth and change that coherence requires. We hope that this book has offered you an experience that has opened your mind to new possibilities that will enrich your relationship with your children as you continue your journey toward integrating a full and coherent life.

# Acknowledgments

Our work together to write this book would never have occurred if not for the support of our family, friends, and colleagues. We are forever grateful for those relationships over the years. Also, we are indebted to the children, parents, students, and teachers who have greatly contributed to our learning about life and relationships.

We thank the parents and faculty at the First Presbyterian Nursery School for their inspiration and their enthusiasm for the blending of the art and science of nurturing at the heart of this book.

We are appreciative of Michael Siegel and Priscilla Cohen for meeting with us during the early phases of writing and for introducing us to our wonderful literary agent, Miriam Altshuler, who has enthusiastically championed this work. We appreciate that our publisher, Jeremy P. Tarcher at Penguin/Putnam, saw the uniqueness of this book and its value for parents. Our editor, Sara Carder, has been a great pleasure to work with as she has helped to sculpt the manuscript and navigate the book through to its final form. Katherine L. Scott's excellent copyediting for the first release further helped smooth out the rough edges. Joanna Ng has been a pleasure to work with in putting together this Tenth Anniversary Edition.

We are grateful for the contributions of Evan Hartzell for his photographic talents and to Mark Pagniano for his artistic expertise.

We wish to thank Mary Main, Ph.D., for her encouragement to make this a scientifically informative book. We also are appreciative of L. Alan Sroufe, Ph.D.'s, review of the final manuscript in its first release, which was beneficial in making the book as current as possible in the areas of attachment and development. We'd also like to thank Deana Eichenstein and Julien Fyrhie for their assistance in researching the latest scientific papers for the Tenth Anniversary Edition.

Along the way we had important feedback from a number of people who were extremely generous of their time and forthright in their insightful comments. We are grateful to Jonathan Fried, Jolee Godino, Lisa Lim, Shelly Pusich, Sarah Steinberg, Melissa Thomas, and Caroline Welch, who helped to make the book more accessible and meaningful for parents.

# About the Authors

Daniel J. Siegel, M.D., received his medical degree from Harvard University and completed his postgraduate medical education at UCLA, where he is currently a clinical professor. The author of the internationally acclaimed *The Developing Mind,* a pioneering book on neurobiology and attachment, and *Brainstorm: The Power and Purpose of the Teenage Brain,* among numerous other books, he is the executive director of the Mindsight Institute, which offers intensive in-person and online educational programs for parents, educators, clinicians, coaches, executives, and scientists. He lives in Los Angeles with his wife and occasionally with his launched adolescents.

Mary Hartzell, M.Ed., received her master's in early childhood education and psychology from UCLA, and is a child development specialist and parent educator. She has taught children, parents, and teachers for more than thirty years and has been the director of a highly respected Reggio-inspired preschool in Santa Monica, California. Her parent-education classes and CD series on parent-child relationships have improved the lives of children and their families for decades.

# Index

Brain stem, 13–14, 71–75, 193–99, 196–97, 202–03
Brenner, J. Douglas, 200–02
Brownell, H., 284–87
Bruner, Jerome, 48–49

Caring communication, 81–82
Categorical emotions, 56–58, 262–63
Central coherence, 283–84
Cerebellum, 77–78, 196–97, 198–200
Cerebral cortex, 196–97, 198–200
Chaos theory, 243–44
Character development, 255–56, 268–69
Childhood amnesia, 12–13, 26–27
Childhood development, 106–07
    communication and, xvi–xix
    parents and, xix–xx
    of personality, xviii–xx
Childhood experiences, 106–07
    and emotions, 61–62
    of parents, 134–37
        and parenting, xv–xvi, xviii–xx, 1–29
Cingulate gyrus, 206–07
Co-construction, 84–85
Cognition, 35–38
Cognitive abilities, xix, xx
Cohen, Donald, 280–82
Coherence, 107–08, 124–25, 140–41, 184–85, 292–93, 294
    central, 283–84
    contingency and, 83–87
Coherent mind, 124–25
Coherent stories, 49–50
    life stories, 39–43, 248, 291–92
Collaborative communication, 67–69, 81–83, 84–85, 90–95, 215–16
    lack of, 95–97
Communication, xvi–xix, 81–104, 232–33, 275–76, 279–80
    breakdowns in, 213–16
    emotional, 53–54, 61–69, 292–94
    in families, 34–35
    integrative, 67–71
    nonverbal, xxiv–xxvi, 59–61, 89–92, 101–103, 124–25, 247–49, 269–70, 283–84
    stories and, 44–45
    See also Collaborative communication; Contingent communication
Compassion, 65–66, 255–58, 260–61
    culture of, 272–74
    development of, xxiv–xxvi
Complexity theory, 249–50
Complex systems, 243–52

Confabulation, 48–50
Connections with others, 91–92, 95–97, 294
    and autonomy, 242–43, 246–50
    positive, 133–34
    ruptures in, 214–18
*Consilience*, Wilson, 20–21
Content of communication, 95–97
Contingent communication, 81–87, 93–94, 98–103, 124–25, 215–16, 291
    and attachment, 107–08
    brain function and, 192–94
    low-mode processing and, 222–23
Contingent connections, 62–63
"Continuous-secure" status, 157–58, 160–61
Convergence zones, 77–78
Corcoran, Rhiannon, 282–83
Corpus Callosum, 13–15, 27–29, 35–38, 77–78, 103–104, 124–25, 198–200, 261–62
Cortical consolidation, 27–29
Cortisol, 121–122
Culture
    of compassion, 272–74
    evolution of, 275–80

Damasio, Antonio, 100–03
Defense mechanisms, psychological, 67–69, 150–52, 225–29, 252
Depression, earned security and, 157–62
*The Developing Mind*, Siegel, 20–21
Devinsky, O., 206–10
Differentiation, 70–71, 249–52
Disconnection, 227–28, 229–30
    in parent-child relationship, 213–16
    shame and, 225–29
Dismissing adult attachment, 139–41, 147–50, 165–69
Disorganized attachment, 110–13, 115–19, 128–29
    unresolved adult attachment and, 136–37, 141–44, 151–53, 162–63
Dissociation, 111–12, 151–52, 180–82
"Double television" experiment, 99–100
Dreams, 27–29
Dyadic regulation, 247–49
Dysregulation, 142–44

Eakin, Paul John, 47–48
Earned-security status, 134–36, 137–40, 156–63
Emotional intelligence, xix–xx, 292–93
Emotions, 22–24, 53–62, 261–63, 291–92
    and adult attachments, 165–68
    and brain, 14–15, 194–96, 232–36
    and communication, 61–64, 90–91, 280–82
    memory and, 11–13

A Sneak Peek at

# Brainstorm

The Power and Purpose of the Teenage Brain

## by Daniel J. Siegel, M.D.

Adolescence is as much a perplexing time of life as it is an amazing one. Running roughly between the ages of twelve and twenty-four (yes, into our mid-twenties!), adolescence is known across cultures as a time of great challenge for both adolescents and the adults who support them. Because it can be so challenging for everyone involved, I hope to offer support to both sides of the generational divide. If you are an adolescent reading this book, it is my hope that it will help you make your way through the at times painful, at other times thrilling personal journey that is adolescence. If you are the parent of an adolescent, or a teacher, a counselor, an athletic coach, or a mentor who works with adolescents, my hope is that these explorations will help you help the adolescent in your life not just survive but thrive through this incredibly formative time.

Let me say from the very start that there are a lot of myths surrounding adolescence that science now clearly shows us are simply not true. And even worse than being wrong, these false beliefs can actually make life more difficult for adolescents and adults alike. So let's bust these myths right now.

One of the most powerful myths surrounding adolescence is that raging hormones cause teenagers to "go mad" or "lose their minds." That's simply false. Hormones do increase during this period, but it is not the hormones that determine what goes on in adolescence. We now know that what adolescents experience is primarily the result of changes in the development of the brain. Knowing about these changes can help life flow more smoothly for you as an adolescent or for you as an adult with adolescents in your world.

Another myth is that adolescence is simply a time of immaturity and teens just need to "grow up." With such a restricted view of the situation, it's no surprise that adolescence is seen as something that everyone just needs to endure, to somehow survive and leave behind with as few battle scars as possible. Yes, being an adolescent can be confusing and terrifying, as so many things during this time are new and often intense. And for adults, what adolescents do may seem confounding and even senseless. Believe me, as the father of two adolescents, I know. The view that adolescence is something we all just need to endure is very limiting. To the contrary, adolescents don't just need to survive adolescence; they can thrive *because* of this important period of their lives. What do I mean by this? A central idea that we'll discuss is that, in very key ways, the "work" of adolescence—the testing of boundaries, the passion to explore what is unknown and exciting—can set the stage for the development of core character traits that will enable adolescents to go on to lead great lives of adventure and purpose.

A third myth is that growing up during adolescence requires moving from dependence on adults to total independence from them. While there *is* a natural and necessary push toward independence from the adults who raised us, adolescents still benefit from relationships with adults. The healthy move to adulthood is toward interdependence, not complete "do-it-yourself" isolation. The nature of the bonds that adolescents have with their parents as attachment figures changes, and friends become more important during this period. Ultimately,

DANIEL J. SIEGEL, M.D.

Taking turns.

we learn to move from needing others' care during childhood, to pushing away from our parents and other adults and leaning more on our peers during adolescence, to then both giving care and receiving help from others. That's interdependence. In this book we'll explore the nature of these attachments and how our need for close relationships continues throughout the life span.

When we get beyond the myths, we are able to see the real truths they mask, and life for adolescents, and the adults in their lives, gets a whole lot better.

Unfortunately, what others believe about us can shape how we see ourselves and how we behave. This is especially true when it comes to teens and how they "receive" commonly held negative attitudes that many adults project (whether directly or indirectly)—that teens are "out of control" or "lazy" or "unfocused." Studies show that when teachers were told that certain students had "limited intelligence," these students performed worse than other students whose teachers were not similarly informed. But when teachers were informed that these same students had exceptional abilities, the students showed marked improvement in their test scores. Adolescents who are absorbing negative messages about who they are and what is expected of them may sink to that level instead of realizing their true potential. As Johann Wolfgang von Goethe wrote, "Treat people as if they were what they ought to be and you help them become what they are capable of being." Adolescence is not a period of being "crazy" or "immature." It is an essential time of emotional intensity, social engagement, and creativity. This is the essence of how we "ought" to be, of what we are capable of, and of what we need as individuals and as a human family.

*Brainstorm* is set up as follows: The first part examines the essence of adolescence and how understanding its important dimensions can create vitality now and throughout the life span. The second part explores the way the brain grows during adolescence so that we can make the most of the opportunities this period of life creates for us. The third part explores how relationships shape our sense of identity and what we can do to create stronger connections with others and with ourselves. In the fourth part, we dive into the ways in which the changes and challenges of adolescence can be best navigated by being present, by being receptive to what is happening so that we can be fully aware of the inner and interpersonal aspects of these experiences. I'll also provide practical steps along the way, in the Mindsight Tools sections, which offer science-proven ways to strengthen our brains and our relationships.

DANIEL J. SIEGEL, M.D.

Because we each learn most effectively in different ways, you may want to experience this book in whatever approach is best for you after reading Part I. If you enjoy learning by blending concepts and facts with science and stories, reading the book from front to back might be best. If instead you learn best by doing, by actual practice, then the four Mindsight Tools sections would be a useful place to start; you can explore the science and stories later. I wrote the book in such a way that if you want to dive into a particular topic, you can do so by reading that part first—for relationships that would be Part III, while for the brain that would be Part II. If you learn best by story-driven discussions, you might read Part IV first, and the earlier sections and practice entries later. Mix it up and find what works best for you. The parts and the Tools sections fit together as a whole; it's up to you how to sample them in a way that meets your needs.

This book is all about understanding and nurturing the essential features of adolescence to bring the most health and happiness into the world regardless of what age we are.

## THE BENEFITS AND CHALLENGES OF ADOLESCENCE

The essential features of adolescence emerge because of healthy, natural changes in the brain. Since the brain influences both our minds and our relationships, knowing about the brain can help us with our inner experience and our social connections. In our journey I'll show how this understanding, and learning the steps to strengthen the brain in practical ways, can help us build a more resilient mind and more rewarding relationships with others.

During the teen years, our minds change in the way we remember, think, reason, focus attention, make decisions, and relate to others. From around age twelve to age twenty-four, there is a burst of growth and maturation taking place as never before in our lives. Understanding the nature of these changes can help us create a more positive and productive life journey.

I'm the father of two adolescents. I also work as a physician in the practice of child, adolescent, and adult psychiatry, helping kids, teens, adults, couples, and families make sense of this exciting time in life. In addition to working as a psychotherapist, I also teach about mental health. What has struck me in each of these roles is that there is no book available that reveals the view that the adolescent period of life is in reality the one with the most power for courage and creativity. Life is on fire when we hit our teens. And these changes are not something to avoid or just get through, but to encourage. *Brainstorm* was born from the need to focus on the positive essence of this period of life for adolescents and for adults.

While the adolescent years may be challenging, the changes in the brain that help support the unique emergence of the adolescent mind can create qualities in us that help not only during our adolescent years, if used wisely, but also as we enter adulthood and live fully as an adult. How we navigate the adolescent years has a direct impact on how we'll live the rest of our lives. Those creative qualities also can help our larger world, offering new insights and innovations that naturally emerge from the push back against the status quo and from the energy of the teen years.

For every new way of thinking and feeling and behaving with its positive potential, there is also a possible downside. Yet there *is* a way to learn how to make the most of the important positive qualities of the teenage mind during adolescence and to use those qualities well in the adult years that come later.

Brain changes during the early teen years set up four qualities of our minds during adolescence: novelty seeking, social engagement, increased emotional intensity, and creative exploration. There are changes in the fundamental circuits of the brain that make the adolescent period different from childhood. These changes affect how teens seek rewards in trying new things, connect with their peers in different ways, feel more intense emotions, and push back on the existing ways of doing things to create new ways of being in the world. Each of these changes is necessary to create the important shifts that happen in our thinking, feeling, interacting, and decision making during our adolescence. Yes, these positive changes have negative possibilities, too. Let's see how each of these four features of the adolescent brain's growth has both upsides and downsides, and how they fill our lives with both benefits and risks.

1. *Novelty seeking* emerges from an increased drive for rewards in the circuits of the adolescent brain that creates the inner motivation to try something new and feel life more fully, creating more engagement in life. **Downside:** Sensation seeking and risk taking that overemphasize the thrill and downplay the risk result in dangerous behaviors and injury. Impulsivity can turn an idea into an action without a pause to reflect on the consequences. **Upside:** Being open to change and living passionately emerge, as the exploration of novelty is honed into a fascination for life and a drive to design new ways of doing things and living with a sense of adventure.

2. *Social engagement* enhances peer connectedness and creates new friendships. **Downside:** Teens isolated from adults and surrounded only by other teens have increased-risk behavior, and the total rejection of adults and adult knowledge and reasoning increases those risks. **Upside:** The drive for social connection leads to the creation of supportive

relationships that are the research-proven best predictors of well-being, longevity, and happiness throughout the life span.

3. *Increased emotional intensity* gives an enhanced vitality to life. **Downside:** Intense emotion may rule the day, leading to impulsivity, moodiness, and extreme, sometimes unhelpful, reactivity. **Upside:** Life lived with emotional intensity can be filled with energy and a sense of vital drive that give an exuberance and zest for being alive on the planet.

4. *Creative exploration* with an expanded sense of consciousness. An adolescent's new conceptual thinking and abstract reasoning allow questioning of the status quo, approaching problems with "out of the box" strategies, the creation of new ideas, and the emergence of innovation. **Downside:** Searching for the meaning of life during the teen years can lead to a crisis of identity, vulnerability to peer pressure, and a lack of direction and purpose. **Upside:** If the mind can hold on to thinking and imagining and perceiving the world in new ways within consciousness, of creatively exploring the spectrum of experiences that are possible, the sense of being in a rut thàt can sometimes pervade adult life can be minimized and instead an experience of the "ordinary being extraordinary" can be cultivated. Not a bad strategy for living a full life!

While we can brainstorm lots of new ideas inside us that we can share collaboratively during the creative explorations and novelty seeking of adolescence, we can also enter another kind of brainstorm as we lose our coordination and balance and our emotions act like a tsunami, flooding us with feelings. That's when we get filled with not only mental excitement but also with mental confusion. Adolescence involves both types of brainstorms.

In a nutshell, the brain changes of adolescence offer both risk and opportunity. How we navigate the waters of adolescence—as young individuals on the journey or as adults walking with them—can help guide the ship that is our life into treacherous places or into exciting adventures. The decision is ours.

Women:
Adult Human Female.